Potato research of tomorrow
Drought tolerance, virus resistance and analytic breeding methods

Mike Difxa

The seminar was organized under the terms of cooperation between the Foundation for Agricultural Plant Breeding SVP, Wageningen, Netherlands, and the Institute for Potato Research Młochów, Rozalin, Poland.

Potato research of tomorrow

Drought tolerance, virus resistance and analytic breeding methods

Proceedings of an international seminar,
Wageningen, Netherlands, 30-31 October 1985

Organizing committee
A.G.B. Beekman, K.M. Louwes, L.M.W. Dellaert and A.E.F. Neele

 Pudoc Wageningen 1986

CIP

ISBN 90 220 0904 1

Printed in the Netherlands.

CONTENTS

PREFACE

The immediate cause for the seminar was the five year cooperation between the Foundation for Agricultural Plant Breeding (SVP) in the Netherlands and the Institute Ziemniaka in Poland. The themes of the seminar, drought tolerance, virus resistance and analytical breeding are subjects of the cooperative research programme.

In the seminar 70 experts participated. The latest results of agronomy, physiology, virology and plant breeding research, which were considered of special importance for the subjects covered by the seminar, were presented and discussed. Through the combination of the various disciplines the multidisciplinary approach in the research is stressed. Besides, posters representing the results of other plant breeding research at the SVP and at the Institute Ziemniaka were shown.

The proceedings are primarily directed to scientists involved in potato research, (potato) breeders and students of agricultural universities. I hope the present work will be of value for the readers, even if I am fully aware of some unavoidable variation in presentation of the results.

Finally, thanks are due to all of the chairmen, authors and participants in the meeting for their contributions and cooperation, and to those who have helped in organizing this seminar.

Lidwine Dellaert,
deputy director SVP
January 15, 1986

Drought tolerance

DROUGHT, A MAJOR CONSTRAINT IN POTATO PRODUCTION AND POSSIBILITIES FOR SCREENING FOR DROUGHT RESISTANCE

C.D. van Loon

Research Station for Arable Farming and Field Production of Vegetables, Lelystad, The Netherlands.

Additional keywords: stress indicators, leaf water potential, osmotic potential, relative water content, stomatal resistance, photosynthesis rate, tuber quality

Summary

The significance of drought as a constraint to world potato production is briefly discussed, as well as the effect of drought on potato growth, production and quality. The following stress parameters and methods for detecting drought stress or drought tolerance are briefly described: leaf water potential, osmotic potential, relative water content, stomatal resistance, photosynthesis rate, water retention by leaves, extent and depth of the root system, anatomical structure of the leaf, a glasshouse wilting and recovery test and field tests. The relationship between some stress parameters and photosynthesis is briefly discussed. Further, varietal differences in reaction to water stress are surveyed. Finally, possibilities for screening clones for drought resistance are discussed. It is concluded that most of the available methods are either insufficiently reliable or too complicated to use on large numbers of clones. The most promising method consists of exposing clones grown in pots in the glasshouse to severe drought. The time taken before wilting symptoms show and the ability to recover after rewatering appear to be well related to drought resistance.

Drought as a constraint to world potato production

Drought is a major constraint to world potato production. Some authors consider it to be the main limiting factor (Van der Zaag & Burton, 1978) influencing negatively not only yield but also tuber quality.

Insufficient water supply may occur almost anywhere potatoes are grown. In arid regions (e.g. sub-tropics), where potato production is only possible with irrigation, short periods of drought often arise because of inadequate irrigation techniques or shortage of water. Even with good irrigation practices water stress may occur, because of high transpiration rates. These rates can be so high, that the root system cannot meet the needs of the plant. This will particularly be the case at low ("dry") water potentials in light soils (Campbell and Kunkel, 1974). Drought may even be a constraint in the humid tropics, in case potatoes can only be grown in the dry season because of diseases such as late blight and brown rot. In the temperate climatic zones both short and long periods of drought may occur in most years, particularly on soils with a low water-holding capacity, such as sands. This causes considerable yield reduction and loss of quality.

Taking into account production conditions and the present yield level, it is estimated that the average potato yield in the world could be increased by at least 50% if the water supply to the crop could be opti-

mized. Breeding for drought resistance should therefore receive high
priority.

Effect of drought on potato growth and production

The effect of drought on the development of the potato plant has been
reviewed extensively by Van Loon (1981). Drought may affect potato
growth and production in three ways: 1) by reducing the amount of productive foliage, 2) by decreasing the rate of photosynthesis per unit of
leaf area and 3) by shortening the vegetative period.

Drought after planting may delay or even inhibit emergence. Gandar and
Tanner (1976) found that leaf elongation of the potato is already
reduced at a mild water stress level, a leaf water potential (LWP) of -3
bars. They found that leaf enlargement ceased completely when the LWP
was lower than -5 bars. Insufficient water supply in the period between
emergence and the beginning of tuber bulking may therefore lead to a
small growth rate of the foliage and to small leaves. This effect will
be even stronger if other growth factors, such as seed quality, are sub-
optimal. For instance, virus-infected seed may give reasonable foliage
development under optimal growth conditions. However, in a dry or poor
soil such degenerate seed will produce plants that remain small. As a
consequence, soil cover with green foliage will often be incomplete and
yields will be below maximum.

Plants under water stress will close their stomata, leading to a
decrease in photosynthesis rate, which results in yield reduction. There
are numerous reports on the effect of drought on potato tuber yields.
See the reviews by Singh (1969) and Harris (1978), from which it can be
derived that for maximum yields soil water content should never be less
than approx. 50% of the water available to the crop during the tuber
bulking period. The critical level of soil water potential has variously
been fixed from 0.2-0.6 bars, depending on climatic, soil and plant
characteristics.

Water stress during the tuber bulking period encourages plant
senescence, resulting in a decreased LAI (Robins & Domingo, 1956;
Susnoschi and Shimsi, 1985). At first the lower leaves start to wilt and
drop off. Simultaneously drought inhibits the development of new leaves.

Drought and tuber quality

Tuber quality characteristics such as shape, dry matter content and
content of reducing sugars can be influenced by water stress during the
vegetative period. Shape defects such as dumb-bell shaped, knobby or
pointed end tubers can be caused by short stress periods during the
tuber bulking period (Steineck, 1958; Nichols and Ruf, 1967). Moorby et
al.(1975) found that these shape defects can be produced after the water
potential of the tubers has been -5 bars for a period of 3 days. Nichols
and Ruf (1967) obtained more malformed tubers from plants stressed for a
period of only 2 days, after which they were watered again, than from
plants under continuous water stress. Growth cracks, another shape
defect, may also be caused by fluctuating water supply (Robins and
Domingo, 1956). It seems that long shaped tubers (e.g. cv. Russet Burbank and cv. White Rose) are more prone to tuber malformation than are
round types (e.g. cv. Kennebec) (Nichols and Ruf, 1967; Wolfe et al.,
1983). Misshapen tubers can also be caused by secondary growth, which is
especially likely to occur in a dry soil with incomplete soil cover by
foliage, when soil temperatures are high (Lugt et al., 1964). This phenomenon may also result in a poor cooking quality (glassiness) among

6

some of the tubers (Van der Zaag, 1958) or in jelly end or translucent end tubers (Iritani and Weller, 1973). All these tubers have a high content of reducing sugars, which causes difficulties for the processing industry. Tubers of water-stressed plants often have a higher content of total sugars than well irrigated plants (Moorby et al., 1975; Levy, 1983 b). Moorby et al. (1975) also found higher content of reducing sugars in tubers from plants subjected to drought, compared with those from control plants; this contrasts to the findings of Wiese et al. (1975), who found low contents of reducing sugars when drought occurred after flowering. Dry matter content is often higher in tubers from water-stressed plants than in well watered plants. Steckel and Gray (1979) found that the dry matter content in potato tubers of 4 varieties grown under low soil moisture was much higher than that of tubers of well-watered plants. These results were confirmed by Shimsi and Susnoschi (1985). Levy (1983 b) however found that tubers from stressed plants of cv. Up to Date and cv. Troubadour had a lower dry matter content and cv. Alpha a higher dry matter content than tubers from well watered plants. Wolfe et al. (1983) found that the dry matter content of tubers was positively affected by only severe water stress: unstressed and moderately stressed plants gave tubers with the same dry matter content.

Drought resistance

According to Levitt (1982), drought resistance can be the result of drought avoidance (e.g. closure of stomata, large root system) or drought tolerance (e.g. capacity for osmotic adjustment, rapid resumption of photosynthesis activity). This terminology will be followed in this paper. Aspects of drought resistance that are considered important and should be taken into account in breeding programmes are: 1) the effect of short periods of water stress on productivity and on tuber quality, 2) survival and recovery of plants after water stress and 3) water use efficiency.

The first aspect has already been discussed. Regarding survival and recovery after water stress, two factors are important: 1) limited loss of soil cover by green leaves and 2) recovery of expansion growth and of the photosynthesis rate. Regarding the second factor, there are indications that the potato is well able to regain its photosynthesis rate after a drought stress period. Chapman and Loomis (1953) found that after "permanently wilted" potato plants were rewatered, the rate of photosynthesis recovered fully, even after 3 days of "permanent wilting". Water use efficiency (WUE), defined as weight of dry matter produced per amount of water (e.g. g.dry matter/kg of water) is particularly important in regions where water is scarce and/or expensive. Compared with other crops the WUE of potatoes is intermediate: 1.56 g. dm/kg of water against 1.92 for barley, 1.85 for wheat and 1.19 for lucerne (Woolhouse, 1983).

Indicators of drought stress or drought resistance

The plant's water status can be described by the following parameters: 1) leaf water potential (LWP) indicating the energy status of the water in the plant and 2) the relative water content (RWC). Plant characteristics such as stomatal resistance and osmotic potential, which are associated with LWP and RWC, can also indicate water stress. If these parameters are related to leaf growth and photosynthesis rates, then they might provide information on drought resistance. Further direct or indirect indication of drought resistance can be obtained by measuring:

- photosynthesis rate
- water retention of excised leaves
- depth and extension of the root system
- anatomical structure of leaf
- yield under dry growing conditions in the field
- the extent of wilting and recovery after severe drought stress.

Leaf water potential

Leaf water potential (LWP) of plants drops as soon as the plant loses more water than it can absorb by its roots. LWP can be measured by e.g. a pressure chamber (Scholander et al., 1965).

Osmotic potential (OP)

Changes in osmotic potential in the cell may be passive (as a result of changes of cell volume as a function of the RWC of the leaf) or active (by accumulation or loss of solutes). OP is directly related to LWP and can be determined e.g. by osmometers (Slavík, 1974).

Relative water content (RWC)

Relative water content, defined as fresh weight minus dry weight / full turgid weight minus dry weight, is a measure of the rate of dehydration of the leaf. The relative water content can be determined by weighing entire leaves or leaf discs immediately after cutting, after prolonged immersion in water, and after oven drying (Turner, 1981).

Stomatal resistance

Plant water stress may lead to the partial or total closure of stomata. As a result, stomatal resistance will increase. This parameter can be measured by diffusive porometers (Slavík, 1974) and by the liquid infiltration technique (Schorn, 1929; Bodlaender, this volume).
For more details about theoretical aspects and measurement of the above mentioned parameters see Vos, this volume.

Photosynthesis rate

Direct measurement of photosynthesis, which has become relatively easy nowadays e.g. by $^{14}CO_2$-pulse equipment (Shimsi and Susnoschi, 1985) or with portable leaf photosynthesis porometers (e.g. LiCor; ADC), is a useful method for detecting water stress in potato plants.

Water retention by leaves

Necas (1974) suggested ascertaining of water retention by leaves as a method of screening for drought resistance. The weight loss of excised potato leaves is recorded under controlled conditions. However, he found a large variability during the growing season and therefore advocated taking measurements for the whole growing season.

Root development

A large root system is thought to be one of the drought resistance mechanisms of plants (Levitt, 1972). Root characteristics that are of importance are rooting depth, the ability of roots to penetrate compacted soil layers, root extension and the growth rate of roots. Rooting depth and extension can be determined by examining soil cores. However, this is a very laborious method. A promising, more rapid alternative could be endoscopy, where root length can be measured by using observation tubes that are placed in the root zone (Vos and Groenwold, 1983).

Anatomical structure of the leaf

Stomatal size and number of stomata per unit area could also be involved in drought resistance, since they influence water loss by transpiration (Levitt, 1972). Also, leaf thickness may differ between stressed and unstressed plants (Levitt, op. cit.).

Yield determination in field experiments

Drought resistance can be assessed in field experiments on soils with a limited water holding capacity. Since replicates are necessary if reliable data are to be obtained, this method is very laborious. This inconvenience can probably partly be overcome by exposing clones to a range of soil moisture conditions by growing them on soil layers of different depths (e.g. 25-85 cm) as suggested by Van der Wal (1977). The soil can be protected from rain by covering the ridges with plastic sheets, in which holes are present to let the plants through.

Extent of wilting and recovery after severe stress

A relatively rapid method to assess the sensitivity of clones to wilting and their ability to survive and to recover from drought is to expose them in pots in the glasshouse to severe stress before rewatering (Beekman, this volume).

Of the above-mentioned factors, LWP, RWC, stomatal resistance and gross photosynthesis rate can be determined rapidly in the laboratory as well as in the field. Determination of osmotic potential and water retention from excised leaves is more laborious. Ascertaining the depth and extent of roots, even if measured with endoscopy, is so time-consuming that it can only be done on a limited scale. The same is true for yield determinations in the field. Exposing plants in pots to severe stress in the glasshouse, although also time-consuming, seems to be a method that allows hundreds of clones to be screened.

Relations between stress parameters

Munns and Pearson (1974) found that under controlled conditions the photosynthesis rate of potatoes was reduced by 20% at an LWP of -5 bars. It seems that the relation between LWP and photosynthesis rate depends on leaf age. Using 5 varieties of potato Susnoschi and Shimsi (1985) found that the gross photosynthesis range (GPR) was, on average, 12% lower in young plants in treatments with a LWP of -10.6 bars compared with plants subjected to an LWP of -8.3 bars. However, in much older plants the rate of GPR was 37% lower in plants with a LWP of -12.1 bars compared with plants with LWP of -9.3 bars. According to Campbell and Kunkel (1974), water stress in the field was not reflected in daytime LWP's, which were as low (-11 bars) as those of well watered plants. However, nocturnal suctions varied between -2 and -3 bars for the well irrigated plants and between -6 and -7 bars for the stressed plants. They suggested that nocturnal LWP's <2 to 3 bars probably cause yield reduction.

Both passive and active osmotic adjustment of the cell sap allow the plant to maintain its photosynthesis rate at a lower LWP (Ludlow, 1981). When the RWC of a leaf decreases, the osmotic value of the cell sap will increase passively. However, it is not clear whether active osmotic adjustment, resulting from water stress, takes place in the potato plant (Vos, this volume).

Decreases in the RWC of the leaf are often associated with a decline in photosynthesis rate (Winkler, 1961; Moorby et al., 1975). Moorby et al. (1975) have found a linear relationship between LWP and RWC.

When stomata close, the stomatal diffusion resistance will increase, resulting in reduced photosynthesis and transpiration. Stomatal resistance of water stressed plants may be in the order of 6 sec cm^{-1} or more, compared with 1-2 sec cm^{-1} for well watered plants (Epstein and Grant, 1973; Campbell and Kunkel, 1974). However, the relationship between stomatal behaviour and plant productivity under stress conditions is not clear. Dwelle et al. (1981) found that correlations between stomatal resistance and yield varied greatly from one season to the next.

Varietal differences in sensitivity to drought

Many reports indicate varietal differences to water stress (Gregersen and Jørgensen, 1973; Steckel and Gray, 1979; Harris, 1978). Roztropowics (1978), comparing 6 varieties, found that without irrigation the yield of the most drought-tolerant variety was 85% of the yield under optimal water supply, whereas the corresponding figure for the least drought-resistant variety was 71%. Shimsi and Susnoschi (1985) did a combined line-source irrigation and nitrogen top-dressing experiment under arid conditions. They found that Spunta, although not the highest yielding variety under optimum water supply, yielded best in the driest rows (table 1).

Table 1. Yield of potato cultivars (t/ha) as a function of levels of irrigation-nitrogen top-dressing, under arid conditions.

	Alpha	Cardinal	Désirée	Spunta	Up to Date
A. Optimum water supply	45.6	57.9	63.6	49.6	54.3
B. Average of the two driest levels	9.9	14.4	15.0	17.1	12.1
B. As % of A	22	25	23.5	34.5	22

(after Shimsi and Susnoschi, 1985)

The 2 varieties that kept their foliage longest under water stress (Alpha and Up to Date) yielded lowest. Shimsi and Susnoschi conclude from these results that the persistance of foliage under water stress is not an indication of the yield potential.

Susnoschi and Shimsi (1985) also found differences between varieties in the partitioning of dry matter when grown under water stress. Cv. Cardinal developed relatively little foliage and accumulated a relatively large part of the dry matter it produced in the tubers. This in contrast to cv. Alpha, which showed the reverse, resulting in a relatively low harvest index.

Root factors

In a pot experiment, Levy (1983 a) found a striking difference in root dry weight between cv. Up to Date and cv. Désirée, the former being 5-7 times heavier than the latter. However, under stress conditions Désirée yielded considerably better. Levy attributed the lower yield of Up to Date to its sensitivity to high temperatures. Steckel and Gray (1979) noticed that in 3 of the 4 years of their experiment the drought-sensitive variety King Edward had a shallower rooting depth than the more drought-resistant Pentland Crown. However, they did not find any relation between rooting depth and drought tolerance for two other varieties.

Iwama et al. (1979) found that in late varieties root dry weight was higher and root diameter was greater than in early varieties. Within the same maturity class, yield differences appeared to be related to differences in root and foliage dry weight. Van Loon (1981) reported on an experiment in which root development of 3 so-called drought-tolerant and 3 drought-sensitive varieties was compared on a soil with and without a ploughpan. At flowering time, the most tolerant variety had the deepest root system and the most sensitive variety had the shallowest root system. Later in the season the differences almost disappeared. None of the varieties penetrated the ploughpan. The latter is in agreement with a conclusion of Pommer (1980), who states that the ability of varieties to penetrate a compacted soil has never been found to be genetically controlled. Van Loon et al. (1985) reported that in a dry year the yield of cv. Bintje with a rooting depth of about 80 cm was more than 50% higher than the yield of the same variety whose rooting depth was only 45 cm because of a ploughpan. From this it can be concluded that there are varietal differences in rooting depth and root weight. However, there is little evidence of a clear relationship between rooting depth and drought resistance. No reports have been found on differences in root activity between varieties.

Anatomical structure of the leaf

Czernik (1978) found a relationship between the structure of leaves and drought tolerance of 2 varieties. The drought-tolerant variety Baca had only half the number of stomata on the underside of the leaf than the less tolerant variety Osa. However, Dwelle et al. (1983) did not find a relation between number of stomata per unit of area and the CO_2 assimilation of potato clones, nor between the latter and the total area of the stomata. Czernik further reported that the intercellular volume of leaves in full turgor from cv. Baca was 10% less than that from cv. Osa.

Stomatal behaviour

There appear to be varietal differences in stomatal behaviour as a reaction to water stress. Levy (1983 a) found that stomata of cv. Up to Date closed earlier under stress than those of cv. Désirée. The latter variety senesced earlier, but gave the highest tuber yield. However, Van Loon (1981) found that under dry conditions the stomata of the relatively drought-tolerant cv. Bintje closed earlier than those of the drought-sensitive cv. Saturna. These contrasting findings would mean

11

that stomatal behaviour is not a suitable indicator of drought resistance. Significant differences between cv. Katahdin and cv. Superior in the pattern of leaf diffusion resistance in response to water stress were reported by Wilcox & Ashley (1980). These results were in agreement with field observations on the drought resistance of these varieties. Dwelle et al. (1983) did not find a relation between stomatal conductance and CO_2 assimilation rates in a group of clones, although these clones differed significantly in stomatal conductance. Vos (this volume) did not find differences in relative stomatal conductance between three cultivars that are known to differ substantially in drought resistance.

These findings suggest that measurement of stomatal behaviour has only limited value for selection for drought resistance.

LWP, osmotic potential and RWC

Levy (1983 a) measured the LWP of stressed and unstressed potato plants of 6 varieties. In unstressed plants differences between varieties were very small (range between -6.3 to -7.5 bars) and in stressed plants they were only slightly greater (range between -7.7 to -9.7 bars). These results were confirmed by Shimsi and Susnoschi (1985), who did not find differences in LWP in unstressed plants among 5 varieties during the growing season. Only in severely water-stressed plants (-13.8 tot -17.7 bars) did some differences develop later in the season.

Small but significant differences in osmotic value of the cell sap between some varieties were found at decreasing soil moisture by Necas (1974); however, it is not clear whether this was the result of passive or active osmotic adjustment. Some other varieties showed differences in osmotic value of the cell sap at high as well as at low soil moisture. Levy (1983 a) reported small varietal differences in osmotic potential of leaves of unstressed (ca -8.5 to -10.- bars) and stressed (ca -9.0 to -11.0 bars) potato plants. The decrease in osmotic potential between stressed and unstressed plants varied between 0.65 and 1.65 bars. However, no indications for osmotic adjustment during stress could be found by Vos (this volume), who criticized Levy's method of ascertaining differences be-tween stressed and unstressed plants. Necas (1974) considered differences in osmotic potential of ca 1.5 bars insufficient to explain any great differences in drought resistance.

Varietal differences in RWC under dry conditions were reported by Werner (1954). In a growth chamber experiment Wilcox & Ashley (1980 and 1982) found only minor differences in RWC between varieties. Epstein and Grant (1973) reported similar results from a field experiment and Vos (this volume) did not find varietal differences in RWC in stressed and unstressed plants either.

Although reports are sometimes contradictory, it can be concluded from the foregoing that neither LWP, OP or RWC can be used to detect drought resistance.

Photosynthesis rate

Shimsi and Susnoschi (1985) measured gross photosynthesis rates of water stressed potato plants and did indeed find differences between varieties, but rates for specific varieties changed considerably during the growing season. Some varieties showed highest rates of CO_2 absorption early in the season and the lowest rates later in the season whereas others showed the reverse. Moreover, there was no relation between gross photosynthesis and yield under conditions of water stress.

Water retention by leaves

Necas (1974) found differences between varieties in their ability to regulate transpiration. Leaves of some varieties lost water more rapidly than leaves from others. Clark and McCaigh (1982) reported similar results with wheat. They found a reasonably good correlation with yields.

Yield determinations on a soil profile with varying depth

Van der Wal (1977) found varietal differences in yield of potatoes growing on soil profiles of different depths and consequently with different amounts of available soil moisture. The results agreed well with the description of these varieties in the Dutch List of Varieties of Arable Crops.

Wilting and recovery after a long water stress period

By exposing several varieties and clones to a long drying cycle in pots in the glasshouse, Beekman (this volume) found large differences between cultivars in the time that they started wilting and in their potential for recovery after rewatering. Since he found a good correlation between these characteristics and the rating for drought resistance in the Dutch List of Varieties of Arable Crops, this method seems to be promising.

Water Use Efficiency (WUE)

Wolfe et al. (1983) did not find differences in WUE between cvs. Kennebec and White Rose. Steckel and Gray (1979) reported that over a period of 3 years King Edward showed a less favourable WUE than other varieties. Versteeg (1985) comparing cv. Revolución and cv. Désirée found a WUE for these cultivars of 3.14 and 3.56 respectively. Bodlaender (this volume) reported a substantial difference in WUE between the drought-resistant cv. Bintje and the drought-sensitive cv. Saturna.

Screening clones for drought resistance

Several methods that could serve as a basis for screening clones for resistance against short drought stress periods have been discussed above. Unfortunately, it must be concluded that none of these methods is both sufficiently reliable and easy to apply, so that it could be used to screen a large number of clones. This conclusion is in agreement with Dwelle (1981) who stated that field measurement of stomatal conductance or stomatal resistance and gross photosynthesis rate have limited value for rapid and extensive selection in potato breeding programmes, since correlations with tuber yield vary greatly between seasons. Also, Shimsi and Susnoschi (1985) come to the conclusion that in the case of water stress there is no clear relationship between leaf permeability, LWP, gross photosynthetic rate and yield under field conditions. This leaves only laborious field methods, including yield determinations, which only allow limited numbers to be screened.

The most promising method for larger numbers of clones seems to be to expose clones grown in pots in the glasshouse to severe drought.

References

*Campbell, G.S. & R. Kunkel, 1974. When to irrigate. p. 63-70. Proc. 13th Annual Washington State Potato Conference & Trade Fair, Moses Lake.

Chapman, H.W. & W.E. Loomis, 1953. Photosynthesis in the potato under field conditions. Plant Physiology 28: 703-716.

Clark, J.M. & Th.N. McCaig, 1982. Evaluation of techniques for screening for drought resistance in wheat. Crop Science 22: 503-506.

Czernik, L., 1978. Relationship between anatomic structure and water deficit in leaves of potato cultivars Baca and Osa. Abstracts of Conference papers 7th Triennial Conference of the EAPR. Warsaw, Poland.

Dwelle, R.B., G.E. Kleinkopf, R.K. Steinhorst, J.J. Pavek & P.J. Hurley, 1981. The influence of physiological processes on tuber yield of potato clones (Solanum tuberosum L.): Stomatal diffusive resistance, stomatal conductance, gross photosynthetic rate, leaf canopy, tissue nutrient levels and tuber enzyme activities. Potato Research 24: 33-47.

Dwelle, R.B., P.J. Hurley & J.J. Pavek, 1983. Photosynthesis and stomatal conductance of potato clones (Solanum tuberosum L.). Plant Physiology 72: 172-176.

Epstein, E. & W.J. Grant, 1973. Water stress relations of the potato plant under field conditions. Agronomy Journal 65: 400-404.

Gandar, P.W. & C.B. Tanner, 1976. Leaf growth, tuber growth and water potential in potatoes. Crop Science 16: 534-538.

Gregersen, A. & V. Jorgensen, 1973. Vanding af Kartofler 1965-71. Irrigation of potatoes. Tidsskrift for Planteavl 77: 611-620.

Harris, P.M., 1978. Water. p. 244-279. In: The potato crop. Ed. P.M. Harris. Chapman and Hall, London.

Iritani, W.M. & L. Weller, 1973. The development of translucent end tubers. American Potato Journal 50: 223-233.

Iwama, K., K. Nakaseko, K. Gotoh, Y. Nishibe & Y. Umemura, 1979. Varietal differences in root system and its relationship with shoot development and tuber yield. Japanese Journal of Crop Science 48 (3): 403-408.

Levitt, J., 1972. Responses of plants to environmental stresses. Academic Press. New York, San Francisco, London. p. 322-446.

Levy, D., 1983 a. Varietal differences in the response of potatoes to repeated short periods of water stress in hot climates. 1. Turgor maintenance and stomatal behaviour. Potato Research 26: 303-313.

Levy, D., 1983 b. Varietal differences in the response of potatoes to repeated short periods of water stress in hot climates. 2. Tuber yield and dry matter accumulation and other tuber properties. Potato Research 26: 315-321.

Loon, C.D. van, 1981. The effect of water stress on potato growth, development and yield. American Potato Journal 58: 51-69.

Loon, C.D. van, L.A.H. de Smet & F.R. Boone, 1985. The effect of a ploughpan in marine loam soils on potato growth. 2. Potato plant responses. Potato Research 28-3 (in preparation).

Ludlow, M.M., 1980. Adaptative significance of stomatal responses to water stress. In: Turner H.C. & P.J. Kramer, Adaptation of plants to water and high temperature stress, Wiley, New York (etc.) p. 123-138.

Lugt, C., K.B.A. Bodlaender & G. Goodijk, 1964. Observations on the induction of second-growth in potato tubers. European Potato Journal 7: 219-227.

Moorby, J., R. Munns & J. Walcott, 1975. Effect of water deficit on photosynthesis and tuber metabolism in potatoes. Australian Journal of Plant Physiology 2: 323-333.

14

Munns, R. & C.J. Pearson, 1974. Effects of water deficit on translocation of carbohydrate in Solanum tuberosum. Australian Journal of Plant
Physiology 1: 529-537.

Necas, J., 1974. Physiological approach to the analysis of some complex
characters of potatoes. Potato Research 17: 3-23.

Nichols, D.F. & R.H. Ruf, Jr. 1967. Relation between moisture stress and
potato tuber development. Proceedings of the American Society for Horticultural Science 91: 443-447.

Pommer, G., 1980. Wurzelwachstum: Umweltabhängigkeit und Möglichkeiten
der züchterischen Beëinflussung. Bayerisches Landwirtschaftliches
Jahrbuch. Heft 8: 936-943.

Robins, J.S. & C.E. Domingo, 1956. Potato yield and tuber shape as
affected by severe soil-moisture deficits and plant spacing. Agronomy
Journal 48: 488-492.

Roztropowicz, S., 1978. Some aspects of Polish physiological and agrotechnical research on the potato. p. 35-60. In: Survey papers 7th
Triennial Conference EAPR, Warsaw, Poland.

Scholander, P.F., H.T. Hammel, E.D. Bradstreet & E.H. Hemmingsen, 1965.
Sap pressure in vascular plants. Science 148: 339-346.

Schorn, M., 1929. Untersuchungen über die Verwendbarkeit der Alkoholfix-
ierungs- und die Infiltrationsmethode zur Messung von Spaltoff-
nungsweiten. Jahrbücher für Wissenschaftliche Botanik 71: 783-840.

Shimsi, D. & M. Susnoschi, 1985. Growth and yield studies of potato
development in a semi arid region. 3. Effect of water stress and
amounts of nitrogen top dressing on physiological indices and on tuber
yield and quality of several cultivars. Potato Research 28: 177-191.

Singh, G., 1969. A review of the soil-moisture relationships in potatoes. American Potato Journal 46: 398-403.

Slavík, B., 1974. Methods of studying plant water relations. Springer
Verlag, Berlin.

Steckel, J.R.A. & D. Gray, 1979. Drought tolerance in potatoes. Journal
of Agricultural Science, Cambridge 92: 375-381.

Steineck, O., 1958. Die Bewasserung der Kartoffel. Deutsche
Landwirtschaftliche Presse 81 N 19: 185-186.

Susnoschi, M. & D. Shimsi, 1985. Growth and yield studies of potato
development in a semi arid region. 2. Effect of water stress and
amounts of nitrogen top dressing on growth of several cultivars.
Potato Research 28: 161-176.

Turner, N.C., 1981. Techniques and experimental approaches for the
measurement of plant water status. Plant and Soil 58: 339-366.

Versteeg, M.N., 1985. Factors influencing the productivity of irrigated
crops in Southern Peru, in relation to prediction by simulation
models. Ph. D. thesis, Agricultural Univeristy, Wageningen.

Vos, J. & J. Groenwold, 1983. Estimation of root densities by observation tubes and endoscope. Plant & Soil 74: 295-300.

Wal, A.F. van der, 1977. Enkele ecologisch-fysiologische aspecten van
onderzoek naar droogte resistentie (unpublished).

Werner, H.O., 1954. Influence of atmospheric and soil moisture conditions on diurnal variation in relative turgidity of potato leaves.
Nebraska Agricultural Experiment Station Research Bulletin 176. 39 p.

Wiese, W., D. Bommer & Chr. Patzold, 1975. Einfluss differenzierter
Wasserversorgung auf Ertragsbildung Knollenqualität der Kartoffel-
pflanze (Solanum tuberosum L.). Potato Research 18: 618-631.

Wilcox, D.A. & R.A. Ashley, 1980. Leaf diffusive resistance and relative
water content as indications of varietal sensitivity to drought in
potatoes. Research Report 62 Storrs Agricultural experiment Station.
The University of Connecticut, Storrs, Connecticut 06268.

Wilcox, D.A. & R.A. Ashley, 1982. The potential use of plant physiological responses to water stress as an indication of varietal sensitivity to drought in four potato varieties. American Potato Journal 59: 533-545.

Winkler, E., 1961. Assimilationsvermögen, Atmung und Erträge der Kartoffelsorten Oberarnbacher Früne, Planet, Lori und Agnes im Tal (610 m) und an der Waldgrenze bei Innsbruck und Vent (1880 m bzw. 2014 m). Flora, Jena 151: 621-661.

Wolfe, D.W., E. Fereres & R.E. Voss, 1983. Growth and yield response of two potato cultivars to various levels of applied water. Irrigation Science 3: 211-222.

Woolhouse, H.W., 1983. The effects of stress on photosynthesis. In: Marcelle, R., H. Clijsters & M. van Poucke (Eds.) Effects of stress on photosynthesis. Nijhoff & Junk, Den Haag, Boston, London.

Zaag, D.E. van der, 1958. Doorwas in aardappelen in 1957. Landbouwvoorlichting 15: 588-599.

Zaag, D.E. van der & W.G. Burton, 1978. Potential yield of the potato crop and its limitations. p. 7-22. In: Survey papers 7th Triennial Conference EAPR, Warsaw, Poland.

RESEARCH ON WATER RELATIONS AND STOMATAL CONDUCTANCE IN POTATOES
1. AN INTRODUCTION TO CONCEPTS, TECHNIQUES AND PROCEDURES.

J. Vos

Centre for Agrobiological Research, Wageningen, the Netherlands

Summary

Introductory descriptions are given of water potential, osmotic potential, pressure potential, and relative water content. The relations
between these parameters are explained with Höfler-Thoday diagrams, with
potatoes and wheat as examples. Definitions are given of the bulk volumetric modulus of elasticity of cell walls and of stomatal resistance and
conductance. The measurement of water potential with the pressure chamber, psychrometers and mechanical press is treated. Osmometers and
different types of porometers are briefly described. Methods to obtain
imprints of the leaf surface for anatomical examination of stomata are
mentioned, as are infiltration techniques. The determinants of plant
water status parameters and of stomatal conductance are briefly discussed.
Keywords: water potential, osmotic potential, osmotic adjustment, osmoregulation, pressure potential, relative water content, pressure chamber,
psychrometer, stomatal resistance, stomatal conductance, potato.

Introduction

Varietal differences in the impact of water stress on yield and quality
of potatoes have been demonstrated (e.g. Steckel & Gray, 1979; Levy,
1983b; Susnoschi & Shimshi, 1985).
It is a working hypothesis that differences in drought resistance
between crop species, or between varieties within species, are partly
attributable to differences in plant water relations and stomatal behaviour. Such hypotheses were tested in many studies (e.g. Turner et al.,
1978; Levy, 1983a; Morgan, 1983; Frank et al., 1984; Markhart, 1985;
Shimshi & Susnoschi, 1985).
In an attempt to contribute to fruitful application of stress physiology in plant breeding, this paper offers an introduction to the basic
theoretical concepts of plant water relations and stomatal behaviour.
Furthermore, techniques and procedures will be discussed. A brief account
is given of the factors and conditions that influence the water status
parameters and stomatal conductance.
The companion paper (this volume) deals with water relations and
stomatal behaviour of three potato varieties that differ in drought
resistance.

Theoretical aspects of water relations and stomatal conductance

Potentials

Commonly the water status in plant systems is measured in terms of
water potential, PSI. This is a measure of the free energy available to
do work, with pure free water as the reference. Water potential is
customarily expressed in pressure units, namely mega pascals or bars

(1 bar = 0.1 MPa) (Dainty, 1976; Jones, 1983; Milburn, 1979).

The main components of water potential that are relevant in plant cells are the osmotic component (PI) and the pressure component (P) so that

$$PSI = PI + P \tag{1}$$

Though derived for cells, eqn 1 is also applied to plant tissues and organs, for instance a leaf.

The pressure potential, or turgor, is the tension in the cell walls brought about by the expansion of the cell contents; the osmotic potential is caused by dissolved substances in the vacuole. The values of P are positive or zero, whereas PI is always negative and PSI negative or zero. Some arbitrarily chosen numerical examples of eqn 1 are:

$$-6 = -10 + 4 \text{ (bar)} \tag{2}$$
$$-6 = - 8 + 2 \tag{3}$$
$$0 = - 6 + 6 \tag{4}$$
$$-10 = -10 + 0 \tag{5}$$

The components of PSI may differ in value between plant tissues or organs, but in equilibrium their sum is constant. Eqns 2 and 3 may, for instance, represent a younger and an older leaf on the same stem, respectively. Eqn 4 applies to a leaf completely saturated with water, whereas eqn 5 shows the situation when turgor is zero.

Relative water content

The relative water content, RWC, is another useful parameter to describe the water status of plant organs. It is defined by:

$$RWC = 100 \ (W_f - W_d)/(W_t - W_d) \tag{6}$$

An alternative expression is the water saturation deficit, WSD, given by:

$$WSD = 100 \ (W_t - W_f)/(W_t - W_d) \tag{7}$$

where in eqns 6 and 7, W_f is the fresh weight upon excision of the organ from the plant, W_t the fully turgid weight after saturation with water, and W_d the dry weight after oven drying.

Höfler-Thoday diagram

The relations between PSI, PI, P and RWC are expressed in the Höfler-Thoday diagram. Two examples, of potato and wheat, are given in Fig. 1. Important characteristics are the values of P and PI at full turgor (PSI = 0, and P = -PI), and the point where turgor reaches zero (P = 0, and PSI = PI).

The elasticity of the cell walls is a property that determines the slopes in the diagram. If the walls are rigid the water potential and its components change rapidly for a given water loss. The bulk volumetric modulus of elasticity, E, is a measure of the elastic properties of the cell walls; it is defined by:

$$E = dP/(dV/V) \text{ (bar or MPa)} \tag{8}$$

where dV/V is the fractional change in volume. E can be approximated by (Jones & Turner, 1978):

18

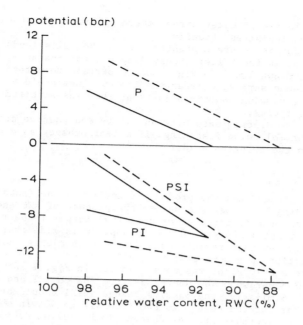

Fig.1. Examples of Höfler-Thoday diagrams. Full drawn lines: fully
   expanded potato leaves (Solanum tuberosum L., cv. Bintje). Broken
   lines: flag leaves of wheat (Triticum aestivum L., cv. Minaret).
   P = pressure potential (turgor), PSI = leaf water potential,
   PI = osmotic potential. Both diagrams pertain to unstressed leaves
   (Vos et al., unpublished).

$$E = 100 \ (dP/dRWC) \tag{9}$$

Fig. 1 shows for wheat leaves lower osmotic potentials and higher
pressure potentials at any RWC than for potato leaves. Furthermore, in
wheat leaves turgor loss occurs at a greater degree of dehydration
(smaller RWC) and at a lower water potential than in potato leaves. Thus,
if the Höfler diagrams of Fig. 1 are representative for both species,
better constitutive traits to cope with drought are indicated for wheat
than for potatoes. However, whether potatoes are generally more drought
sensitive than wheat cannot be determined with Höfler-Thoday diagrams
only. This also depends on factors like root distribution, resistances to
water flow in the plant and water use efficiency (kg water transpired per
kg dry matter produced).

Osmotic adjustment

When a leaf dehydrates the osmotic potential declines passively,
because the concentration of solutes in the cells increases. Lowering of
the osmotic potential by active accumulation of solutes in the cells in
response to water stress has been observed in several plant species. This
process is called osmotic adjustment (Turner & Jones, 1980; Morgan, 1980)
or osmoregulation (Morgan, 1983; Kumar et al., 1984). The significance of
osmotic adjustment is that turgor and turgor-dependent processes are

19

maintained at levels of water stress where otherwise plant functioning would have been inhibited strongly.

Active increase in solute concentration in the cells leads to changes in the relations in the Höfler-Thoday diagram. For instance, hypothetically the full drawn lines in Fig. 1 could pertain to potato leaves that did not experience stress, whereas the broken lines could pertain to similar leaves, showing osmotic adjustment that was acquired during one or more drying cycles.

An other way to discriminate between active and passive changes in PI is based on the following reasoning: if a leaf behaves as a perfect osmometer then the relation holds:

$$RWC_x . PI_x = RWC_o . PI_o \qquad\qquad (10)$$

This equation states that the product of any RWC value (subscript x) and the corresponding PI value is equal to the product of RWC and PI at a reference point, for instance at or near full turgor (subscript o) (law of Boyle-van 't Hoff). In eqn 10 the assumption is made that the fraction of total leaf water that resides outside the plasmalemma (i.e. apoplastic water, see section on osmometers) is small.

According to eqn 10 PI of the potato leaves in Fig. 1 should fall from −7.3 bar to −7.8 bar when RWC decreases from 98% to 92% and only passive changes occur. Similarly, the passive change in PI should be from −10.8 bar to −11.6 bar for the change in RWC from 97% to 90% in wheat leaves. Fig. 1 shows considerable greater changes in PI during dehydration. This leads to the conclusion that active accumulation of solutes occurred in both species. However, the data in Fig. 1 pertain to unstressed leaves and therefore the alternative conclusion is that the conditions that justify the application of eqn 10 are not met. For instance, if the osmotic potential of xylem fluids is not close to zero at full turgor and declines during dehydration, eqn 10 does not hold. P.M. Schildwacht (State University, Utrecht; personal communication, 1985) indeed observed osmotic potentials in xylem sap of potato leaves of several bars, while the values apparently depended on the degree of dehydration of the plant in the course of the day.

Morgan (1983; see also Kumar et al., 1984) extended the analysis of the relation between PI and RWC. Eqn 10 can be rearranged and brought in a linear form by taking the natural logarithm to obtain:

$$lnPI_x = ln(RWC_o . PI_o) - b \; lnRWC_x \qquad\qquad (11)$$

Regression analyses can be made of lnPI on lnRWC (use absolute values of PI). If only passive changes of PI occur, the slope of eqn 11, b, equals 1, whereas smaller values indicate active accumulation of solutes.

Stomatal resistance or conductance

Analogous to Ohm's law, transport rates in the soil-plant-atmosphere continuum are commonly described in terms of driving forces and resistances to transfer. Transpiration, that is the transport of water from the leaf interior to the atmosphere, can also be described in such terms.

The stomata constitute a major resistance to transpiration. Moreover, plants are able to regulate stomatal opening in response to internal and external conditions. Therefore stomatal resistance is a relevant parameter to measure, especially in studies on the effect of water stress.

The most frequently used definition of stomatal resistance, SR, is:

SR = water vapour density gradient/transpiration rate (12)

The water vapour density gradient is expressed in g cm^{-3}, the transpiration rate in g $cm^{-2} s^{-1}$, and so SR is expressed in s cm^{-1}.

Instead of stomatal resistance it is often more convenient to use stomatal conductance, SC, as defined by:

$$SC = 1/SR \ (cm \ s^{-1})$$ (13)

The water vapour density gradient is calculated as the difference in density in the sensor cuvette and in the leaf interior, where the latter is set equal to the saturated density of pure water at the prevailing temperature of the leaf.

In eqn 12 the resistance is the sum of a series of resistances (ignoring cuticular resistance because it is large). The resistance to transfer through the laminar air layer at the leaf surface (boundary layer resistance) can be measured and accounted for, leaving the "true" stomatal resistance as the main determinant of the remaining resistances. The boundary layer resistance is small compared to the stomatal resistance (see e.g. Jones, 1983).

Methods, equipment and procedures

Measurement of water potential

The main instruments that can be used to determine water potential are the pressure chamber, thermocouple psychrometers, and the "J-14" mechanical press.

A pressure chamber (Scholander et al., 1965) consists of a cilinder with a removable lid. The chamber can be pressurized, using a cilinder of compressed air or nitrogen as a source. A pressure gauge shows the pressure in the system. A hole or slit is drilled in the lid. The severed end of a petiole, a leaf blade, a stolon or any other organ is fixed in a rubber bung (or other material, see e.g. Turner, 1981; Brown & Tanner, 1981) provided with a hole or a slit to accommodate the organ in a manner causing neither air leaks, nor crushing of the plant tissue. The chamber is closed, with the cut end of the organ just protruding from the chamber. Subsequently the chamber is pressurized at a rate of 3 bar min^{-1}, or less (Turner, 1981). When the sap returns to the severed ends of the xylem vessels the pressure (= water potential) is recorded; this is called the "endpoint". A hand lens with a magnifying factor of six to ten is needed to observe the endpoint.

It is a good practice to cover leaves with a plastic sheath from the time of excision from the plant till completion of the measurement (Gandar & Tanner, 1976a; Wenkert et al., 1978).

Depending on the additional work an operator has to do (e.g. weighing of leaves, etc.) 10 to 30 measurements can be made per hour.

The advantages of the pressure chamber technique are that a straightforward measurement of water potential is obtained, while the instrument is relatively simple to operate, also in the field. For the novice it can be a difficulty to learn to distinguish between "false" and "true" endpoints. False endpoints can occur when gas escapes through the intercellular spaces, causing vigorous bubbling on the cut surface. Touching the cut end with paper tissue usually helps. The endpoint can be checked by lowering the pressure in the chamber by one to two bar and subsequent repressurization (Brown & Tanner, 1981).

The thermocouple psychrometer is based on the principle that the relative water vapour pressure (e/e_o) of a solution or a piece of plant material is related to its water potential, PSI, according to:

$$PSI = (RT/\bar{V}) \ln(e/e_o) \quad (Pa) \tag{14}$$

were R is the gas constant (8.3144 J mol^{-1} T^{-1} or Pa m^3 mol^{-1} T^{-1}), T the Kelvin temperature, $\bar{V}$ the partial molal volume of water (18.048 10^{-6} m^3 mol^{-1} at 20 °C), e the equilibrium vapour pressure of water (Pa) above the solution or the plant tissue, and e_o the saturated vapour pressure of free water at atmospheric pressure. For the expression in bar the result of eqn 14 has to be multiplied by 10^{-5}. The instrument can be calibrated with salt solutions, of which the relationship between e and the concentration of the solution is known.

Psychrometers can be operated in the psychrometric mode (wet bulb depression; Turner et al., 1984) or in the dew point mode (Brown and Tanner, 1981; Savage et al., 1983). Psychrometers can be used in situ (Savage et al., 1983; Shackel, 1984) or with leaf discs (Roy & Berger, 1983). Several comparisons between psychrometry and pressure chamber were made (e.g. Brown & Tanner, 1983; Turner et al., 1984).

Although various research workers successfully employed psychrometry, it is at present not a technique that can be recommended for routine use in applied research.

A relatively new technique is the measurement of water potential with the "J-14" hydraulic press, manufactured by Campbell Scientific Inc., Logan, Utah, USA. A leaf or a disc is placed in the chamber between a flexible membrane and a transparent lid. The pressure, provided by a hydraulic pump, squeezes the plant tissue between the flexible membrane and the transparant window. The pressure required to induce a colour change of the leaf and to express water at the petiole or at the cut edge is taken as a measure of water potential. The apparatus is cheap and simple. It can be used in the field and the measurement is rapid. The determination of the endpoint can be subjective. The apparatus can also be used for the determination of soil water potential. Hunt et al. (1984) found that the regression equations between pressure chamber measurements and J-14 readings differed for various tree species and were presumably dependent on the rigidity of the structural material of the leaves. Rajendrudu et al. (1983) concluded that for groundnuts the performances of the J-14 press, a dew point psychrometer and the pressure chamber were similar.

The press seems to offer promise in applied research, but calibration against a standard technique is necessary.

Osmometers

Osmotic potential can be determined in various ways (Slavík, 1974). Most commonly used are osmometers, based on the principle of freezing point depression or lowering of the equilibrium water vapour pressure above solutions relative to free water. For both types of instruments calibration with salt solutions of known activity is necessary. The osmotic potential of salt solutions can be calculated according to:

$$PI = aRTc \quad (Pa) \tag{15}$$

were a is the activity coefficient of the salt solution (accounting for non-ideal osmotic behaviour), which is dependent on the concentration and is tabulated in handbooks, R is the gas constant, T the Kelvin temperature, and c the concentration of the solution in mmol per kg water, accounting for the number of osmotically active particles per molecule. For the expression in bar, the result of eqn 15 has to be multiplied by 10^{-5}.

For the determination of osmotic potential samples have to be frozen immediately upon removal from the plant or after measurement of water potential. To achieve this the sample can be put in a small closed plastic container that is immersed in liquid nitrogen (-196 °C) (see also Slavík, 1974 and Turner, 1981). Subsequently, samples are stored at -20 °C. Upon thawing, sap can be expressed with a hand-operated kitchen press or other equipment designed to separate the structural material from the sap (e.g. Turner, 1981). An aliquot of the sap is immediately measured in the osmometer.

After freezing and thawing the cell integrity is destroyed, causing water outside the cell plasmalemma (apoplastic water) to mix with water inside the plasmalemma (symplastic water). Because the osmotic potential of apoplastic water is usually close to zero the measured osmotic potential is somewhat higher (less negative) than in the intact cell. The magnitude of this dilution can be assessed from the relationship between PSI and PI below the point of turgor loss (Roy & Berger, 1983; and Fig. 3 companion paper). Furthermore, the apoplastic water fraction can be estimated from pressure-volume curves (see a later section).

The determination of turgor

Pressure potential or turgor is generally calculated as the difference between the measured water potential and the measured osmotic potential (eqn 1). Nowadays it is possible to measure cell turgor, cell wall elasticity and related parameters directly with a microcapillary connected to a pressure probe (Zimmermann & Hüsken, 1979). However, the technique is not suitable for routine application in applied research.

The determination of relative water content

Entire leaves or leaf discs can be used for the determination of their relative water content. It follows from eqn 6 that this determination is a matter of weighing the leaves or discs (10 to 15 of about 1 cm² area per sample) immediately after excision, after saturation, and after oven drying.

To saturate the discs they are floated on distilled water in covered petri dishes, or fitted in holes in a wet polyurethane foam sheet (Slavík, 1974; Turner, 1981). When entire leaves are used the petiole is placed in a cuvette with water.

During a short initial period (hours or less) water uptake is very rapid; subsequent water uptake is slow, but can continue for a long period (24 hours or more). This continuous slow uptake of water can be due to expansion growth, but often full grown leaves also exhibit continuous slow water uptake over a period of 24 hours or more. If the saturation period is taken too long, the curve of RWC against PSI will show a large intersect with the RWC axis.

For potatoes we adopted the following procedure. Upon completion of the PSI measurement (pressure chamber) the leaves were immediately weighed to the nearest 0.001 g on an electronic balance. (The balance was encased in a wooden box (with a window in the lid) to prevent the impact of wind on the weighing.) After weighing, the leaves were put into small plastic cuvettes with their petioles in water. The cuvettes were placed in a

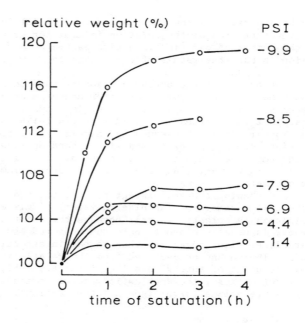

Fig. 2. Change with time of relative fresh weight of excised potato
leaves with their petioles in water; initial leaf water potential
(PSI) at time zero varied between -1.4 and -9.9 bar. Each curve
represents a single leaf. See text for further details. The data
are from the experiment described in the companion paper in this
volume.

closed, dark box. Water uptake tests (Fig. 2) and PSI-RWC plots showed
that three hours was an adequate time lapse for the saturation of leaves
varying in initial water potential between -1 and -10 bar.

Pressure-volume curves

Pressure chambers can be used to determine the relationship between PSI
and RWC of single leaves. The procedure is started with a fully turgid
leaf. Turgid leaves can be obtained from well watered plants before dawn,
or from plants that were covered the night before; alternatively the
leaves can be saturated as in the determination of RWC. The leaves are
stepwise dehydrated by over-pressurization in the pressure chamber.
During this process PSI is measured as a function of water loss. Water
loss can be measured by collecting the extruded sap or by intermediate
weighings. The dehydration process is continued till PSI is -20 to -40
bar. After termination of the dehydration process, the oven dry weight of
the leaf is determined.
    The initial data consist of an array of PSI values and the corres-
ponding cumulative water loss. The data are analysed by plotting the
reciprocal of the balance pressure (the pressure in the pressure chamber
at each endpoint) against RWC or against the amount of water that is
lost. From such plots inferences can be made on the same parameters that
can be derived from a Höfler-Thoday diagram (Fig. 1, based on multiple
determinations on many leaves). The osmotic potentials at full and zero
turgor can be calculated without direct measurement of these parameters.

Furthermore, the bulk volumetric modulus of elasticity and the apoplastic water fraction can be estimated. Relevant literature reports are those by Hellkvist et al. (1974), Richter (1978), Turner (1981), Brown & Tanner (1983) and Ritchie & Roden (1985).

The main disadvantage of the method is its time requirement (Ritchie & Roden, 1985).

The use of the pressure-volume technique should be considered if one wants to examine possible differences in water relations between species or cultivars (Frank et al., 1984).

Measurement of stomatal functioning

The most physiologically meaningful measure of stomatal functioning is the diffusion resistance or conductance to water vapour. This parameter can be determined with two types of diffusion porometers: transit-time and continuous flow (= steady state) porometers (Jones, 1983; Slavík, 1974). Transit-time porometers measure the increment of the humidity during a certain time lapse in a closed sensor cuvette that is attached to a leaf surface. From this reading stomatal conductance can be worked out. With continuous flow porometers the transpiration rate of a leaf portion is derived from the flow rate of dry air that has to be passed along the leaf portion in order to keep the relative humidity of the outgoing air at a preset level. Stomatal conductance is calculated as given by eqns 12 and 13.

There are other types of porometers (Jones, 1983; Slavík, 1974; Moreshet & Falkenflug, 1978; Shimshi et al., 1983) that provide an indirect measure of conductance, based on diffusion, or mass flow under a pressure gradient, of inert or radio-active gas. Especially when trans-leaf measurements are made the results should be interpreted with care. For the latter type of measurement leaf permeability is perhaps a more suitable term than conductance (Shimshi et al., 1983). Dwelle et al. (1981, 1983) used a porometer where the diffusion of tritiated water into the leaf was taken as a measure of conductance.

Graded solutions of differing viscosity (e.g. various mixtures of liquid paraffin and kerosine or xylene) are used to estimate relative stomatal opening. The viscosity of the solution that just infiltrates the leaf (indicated by a change in leaf colour) within a fixed number of seconds provides a measure of aperture (Slavík, 1974; see also Bodlaender this volume).

Measurement of stomatal size and frequency

Measurement of stomatal size and frequency can be made either directly or on imprints. With the latter technique cellulose acetate (Dwelle et al., 1983) or nail polish is spread over the leaf surface. Upon hardening the material is peeled off and examined under a microscope. In a two-step procedure an imprint is made with xantopren (Bayer Dental) from which a leaf replica is made with polystryrene, dissolved in toluol.

Determinants of plant water status and stomatal conductance

In this section a brief and qualitative account will be given of factors and conditions that affect the parameters of the water status and stomatal conductance (see also Van Loon, this volume).

Water potential, osmotic potential, pressure potential and relative water content change with the degree of hydration, as expressed by the Höfler-Thoday diagram (Fig. 1).

Even with optimal water supply plants dehydrate during the day, and so potential gradients are created that act as driving forces for water transfer from the soil to the atmosphere. The degree of dehydration depends upon the evaporative demand (which, in turn, is a function of the radiation load and the vapour pressure deficit of the air) and on the magnitude of the resistances to flow in the soil-plant-atmosphere continuum.

With sub-optimal supply of water, plants dehydrate faster after sunrise and often to a greater extent. However, stomatal closure can be induced by water stress. In this way further desiccation as well as photosynthesis are reduced (Bodlaender, this volume; Vos et al., in preparation). Stomatal closure is indicated by decreasing stomatal conductance. It can be mentioned in passing that, when water stress occurs, expansion growth is reduced because leaf water potential (or turgor) is below the threshold for expansion for a longer fraction of the 24-hour period (Gandar & Tanner, 1976b).

Stomatal conductance also depends on factors as radiation intensity (Dwelle et al., 1983), (internal) carbon dioxide concentration and air humidity (c.f. Jones, 1983).

Both maximal stomatal conductance and the water relations (Höfler-Thoday diagram) change with leaf age; the decline in maximum conductance is correlated with the decline in maximum photosynthetic capacity with leaf age (Vos et al., in preparation).

It follows from the foregoing that in research on the impact of water stress, determinations of the water status and of stomatal conductance can be interpreted only when the measurements are made on specified leaves with measurements on controls, receiving optimal water, serving as a reference. This statement also applies when plant parameters are used to assess the need for irrigation. The possible significance of the determination of plant water relations and stomatal behaviour to identify parental lines or progenies with a high degree of drought resistance will be discussed in the companion paper (see also Beekman, this volume).

References

Brown, P.W. & C.B. Tanner, 1981. Alfalfa water potential measurement: a comparison of the pressure chamber and leaf dew-point hygrometers. Crop Science 21: 240-244.

Brown, P.W. & C.B. Tanner, 1983. Alfalfa osmotic potential: a comparison of the water-release curve and frozen-tissue methods. Agron. J. 75: 91-93.

Dainty, J., 1976. Water relations of plant cells. In: U. Lüttge & M.G. Pitman (Eds.): Transport in plants II, part A cells; Encyclopedia of Plant Physiology New Series Volume 2. Springer-Verlag, Berlin. p. 13-16.

Dwelle, R.B., G.E. Kleinkopf, R.K. Steinhorst, J.J. Pavek & P.J. Hurley, 1981. The influence of physiological processes on tuber yield of potato clones (Solanum tuberosum L.): stomatal diffusive resistance, stomatal conductance, gross photosynthetic rate, leaf canopy, tissue nutrient levels, and tuber enzyme activities. Potato Res. 24: 33-47.

Dwelle, R.B., P.J. Hurley & J.J. Pavek, 1983. Photosynthesis and stomatal conductance of potato clones (Solanum tuberosum L.). Comparative differences in diurnal patterns, response to light levels, and assimilation through upper and lower leaf surfaces. Plant Physiol. 72: 172-176.

Frank, A.B., R.E. Barker & J.D. Berdahl, 1984. Pressure-volume characteristics of genotypes of three wheatgrass species. Crop Science 24 (2): 217-220.

Gandar, P.W. & C.B. Tanner, 1976a. Potato leaf and tuber water potential measurements with a pressure chamber. Am. Pot. J. 53: 1-14.

Gandar, P.W. & C.B. Tanner, 1976b. Leaf growth, tuber growth, and water potential in potatoes. Crop Science 16: 534-538.

Hellkvist, J., G.P. Richards & P.G. Jarvis, 1974. Vertical gradients of water potential and tissue water relations in sitka spruce trees measured with the pressure chamber. J. appl. Ecol. 11: 637-667.

Hunt, E.R., J.A. Weber & D.M. Gates, 1984. Differences between tree species in hydraulic press calibration of leaf water potential are correlated with specific leaf area. Plant, Cell and Environment 7: 597-600.

Kumar, A., P. Singh, D.P. Singh, H. Singh & H.C. Sharma, 1984. Differences in osmoregulation in Brassica species. Ann. Bot. 54: 537-541.

Levy, D., 1983a. Varietal differences in the response of potatoes to repeated short periods of water stress in hot climates. 1. Turgor maintenance and stomatal behaviour. Potato Res. 26: 303-313.

Levy, D., 1983b. Varietal differences in the response of potatoes to repeated short periods of water stress in hot climates. 2. Tuber yield and dry matter accumulation and other tuber properties. Potato Res. 26: 315-321.

Jones, H.G., 1983. Plants and microclimate. A quantitative approach to environmental physiology. Cambridge University Press, Cambridge.

Jones, M.M. & N.C. Turner, 1978. Osmotic adjustment in leaves of sorghum in response to water deficits. Plant Physiol. 61: 122-126

Markhart, A.H., 1985. Comparative water relations of Phaseolus vulgaris L. and Phaseolus acutifolius Gray. Plant Physiol. 77: 113-117.

Milburn, J.A., 1979. Water flow in plants. Longman, London.

Moreshet, S. & V. Falkenflug, 1978. A krypton diffusion porometer for the direct measurement of stomatal resistance. J. exp. Bot. 29 (108): 267-275.

Morgan, J.M., 1980. Differences in adaptation to water stress within crop species. In: N.C. Turner & P.J. Kramer (Eds): Adaptation of plants to water and high temperature stress. John Wiley & Sons, New York. p. 369-382.

Morgan, J.M., 1983. Osmoregulation as a selection criterion for drought tolerance in wheat. Aust. J. agric. Res. 34: 607-614.

Rajendrudu, G., M. Singh & J.H. Williams, 1983. Hydraulic press measurements of leaf water potential in groundnuts. Expl. Agric. 19: 287-291.

Richter, H., 1978. A diagram for the description of water relations in plant cells and organs. J. exp. Bot. 29 (112): 1197-1203.

Ritchie, G.A. & J.R. Roden, 1985. Comparison between two methods of generating pressure-volume curves. Plant, Cell and Environment 8: 49-53.

Roy, J. & A. Berger, 1983. Water potential measurement, water compartmentation and water flow in Daytylus glomerata L. leaves. New Phytol. 93: 43-52.

Savage, M.J., H.H. Wiebe & A. Cass, 1983. In situ field measurement of leaf water potential using thermocouple psychrometers. Plant Physiol. 73: 609-613.

Scholander, P.F., H.T. Hammel, E.D. Bradstreet & E.A. Hemmingsen, 1965. Sap pressure in vascular plants. Science 148: 330-346.

Shackel, K.A., 1984. Theoretical and experimental errors for in situ measurements of plant water potential. Plant Physiol. 75: 766-772.

Shimshi, D., J. Shalhevet & T. Meir, 1983. Irrigation regime effects on some phsiological responses of potato. Agron. J. 75: 262-267.

Shimshi, D. & M. Susnoschi, 1985. Growth and yield studies of potato development in a semi-arid region. 3. Effects of water stress and amounts of nitrogen top dressing on physiological indices and on tuber yield and quality of several cultivars. Potato Res. 28: 177-191.

Slavík, B., 1974. Methods of studying plant water relations. Springer-Verlag, Berlin.

Steckel, J.R.A. & D. Gray, 1979. Drought tolerance in potatoes. J. agric. Sci., Camb. 92: 375-381.

Susnoschi, M. & D. Shimshi, 1985. Growth and yield studies of potato development in a semi-arid region. 2. Effect of water stress and amounts of nitrogen top dressing on growth of several cultivars. Potato Res. 28: 161-176.

Turner, N.C., 1981. Techniques and experimental approaches for the measurement of plant water status. Plant and Soil 58: 339-366.

Turner, N.C., J.E. Begg, H.M. Rawson, S.D. English & A.B. Hearn, 1978. Agronomic and physiological responses of soybean and sorghum crops to water deficits. 3. Components of leaf water potential, leaf conductance, $^{14}CO_2$ Photosynthesis, and adaptation to water deficits. Aust. J. Plant Physiol. 5: 179-194.

Turner, N.C. & M.M. Jones, 1980. Turgor maintenance by osmotic adjustment: a review and evaluation. In: N.C. Turner & P.J. Kramer (Eds): Adaptation of plants to water and high temperature stress. John Wiley & Sons, New York. p. 87-103.

Turner, N.C., R.A. Spurway & E.-D. Schulze, 1984. Comparison of water potentials measured by in situ psychrometry and pressure chamber in morphologically different species. Plant Physiol. 74: 316-319.

Wenkert, W., E.R. Lemon & T.R. Sinclair, 1978. Changes in water potential during pressure bomb measurement. Agron. J. 70: 353-355.

Zimmermann, U. & D. Hüsken, 1979. Theoretical and experimental exclusion of errors in the determination of the elasticity and water transport parameters of plant cells by the pressure probe technique. Plant Physiol. 64: 18-24.

RESEARCH ON WATER RELATIONS AND STOMATAL CONDUCTANCE IN POTATOES
2. A COMPARISON OF THREE VARIETIES DIFFERING IN DROUGHT TOLERANCE

J. Vos

Centre for Agrobiological Research, Wageningen, the Netherlands

Summary

Water relations and stomatal conductance during transient water stress
of the cvs Kennebec, Bintje and Saturna were studied in an experiment, in
which the plants were grown on large containers under a rain shelter.
Controls received water throughout the growing period. Kennebec is
supposed to be the most and Saturna the least drought tolerant of these
cultivars.
Before and during water stress, as well as after its relief, determinations were made of leaf water potential, osmotic potential and relative
water content on fourth and fifth leaves from the top. Measurements of
stomatal conductance were made on the same leaves with a Li-Cor 1600
steady state porometer.
For most of the growing period the control of Bintje showed about 10%
smaller values of stomatal conductance. Relative stomatal conductance,
RSC, of stress treatments declined rapidly when soil water became
limiting. Beyond five days after relief of stress RSC exceeded 100%.
The relations between leaf water potential and relative water content,
as between osmotic potential and water potential were not affected by
variety, time of measurement, and stress treatment. It is concluded that
under the conditions of the experiment the varieties tested did not
differentiate with respect to water relations, neither with optimal
supply of water nor during water stress. There were no indications of
active lowering of the osmotic potential in response to stress. The
pattern of change in stomatal conductance during transient water stress
was also similar for the varieties.
Keywords: water potential, osmotic potential, relative water content,
stomatal conductance, stomatal resistance, potato, variety, drought.

Introduction

It is a working hypothesis that varietal differences in drought tolerance are partly attributable to differences in water relations and in
stomatal behaviour (Levy, 1983).
Differences in water relations can consist of, or arise from:
a. a lower osmotic potential at full turgor
b. a lower leaf water potential and relative water content at the point
   where turgor reaches zero
c. differences in the bulk volumetric modulus of elasticity
d. active lowering of the osmotic potential in response to water stress
   (Turner & Jones, 1980; Morgan, 1980).
Varietal differences in stomatal conductance under stress and non-
stress conditions were reported by Levy (1983), whereas Dwelle et al.
(1981, 1983) showed such differences in the absence of water stress.
In this paper preliminary results are reported of a study on the impact
of transient water stress on leaf water relations and stomatal behaviour
of three varieties that are known to differ substantially in drought

sensitivity, namely Kennebec, Bintje and Saturna. According to tests made at the Foundation for Plant Breeding in Wageningen (Beekman, this volume), Kennebec is the most and Saturna the least drought tolerant of these cultivars. Furthermore, in the Dutch List of Varieties of agricultural Crops (Commissie voor de Samenstelling van de Rassenlijst voor Landbouwgewassen, 1985) Kennebec is classified in the category of varieties with the highest degree of drought tolerance, Bintje in the second highest, and Saturna in the category with the lowest tolerance. Further criteria to select these varieties were a similar category of earliness, and diversity in genetic background: Kennebec was bred in the USA, whereas Bintje and Saturna are Dutch cultivars.

There were also data collected on growth and development, mineral uptake, and water use efficiency, but a more extensive account of this work will be published elsewhere.

Material and Methods

Potatoes were grown under an open rain shelter on containers with dimensions 1.50 X 1.33 m, and a soil depth of 0.35 m. A wooden frame, covered with nylon gauze was placed on the bottom of the containers. The space underneath the soil layer served as a water reservoir. When supplying water the tap was turned off when water started to run from an overflow at 2 cm height above the bottom of the soil layer.

On 18 April 1985 six seed potatoes were planted in each of two rows; the distance between rows was 0.66 m. Controls received water throughout the growing period. On 14 June (56 DAP (days after planting)) the water underneath the soil layer was drained for part of the containers; in this way drought was gradually induced. On 90 DAP watering was resumed.

Before and during stress, as after its relief, stomatal behaviour and water relations were measured. The fourth and fifth leaves from the top were selected for these measurements. Certainly on the earlier sampling days these leaves were still expanding. Measurements on water relations were made from dawn till late in the afternoon. Stomatal conductance was usually measured between 2.5 hours before and after solar noon.

Immediately upon removal from the plant leaf water potential was determined with a pressure chamber. Subsequently, the leaves were alternately subjected to measurement of relative water content or osmotic potential (for procedures see companion paper). Osmotic potentials were determined with a Wescor model 5500 vapour pressure osmometer. Stomatal conductance was measured with a Li-Cor 1600 steady state porometer; unless stated otherwise the data pertain to the lower leaf surface.

Results

Stomatal conductance of controls

Table 1 provides a compilation of stomatal conductance (SC) in controls of the three varieties. The data were selected to demonstrate some factors and conditions that affect the value of SC. On bright days with a high evaporative demand partial closure of stomata can occur. This is shown by the lower values of SC in the afternoon than in the morning on 77 DAP. The data from 82 and 83 DAP are included to shown that SC is lower on days with low radiation levels. The data collected on days with fairly comparable weather conditions show that SC declined with age of the crop.

Lastly, Bintje showed SC values that were about 10% lower than those of Kennebec and Saturna, except after about 100 DAP when Saturna appeared to senesce faster. In one test SC of upper and lower leaf surfaces were compared. SC of upper surfaces was expressed in per cent of SC of lower surfaces; the mean values were 9% for Bintje and 4% for both Kennebec and Saturna.

Table 1. A compilation of stomatal conductance (cm s^{-1}) of control treatments (n = 15 to 36)

Days after planting	Weather conditions	Time of day of measurement (h)	Variety		
			Kennebec	Bintje	Saturna
77	bright	10.30 – 12.15	1.53	1.37	1.65
77	bright	14.00 – 16.30	1.16	0.87	0.87
78	sunny	9.15 – 15.00	1.48	1.34	1.49
82, 83	cloudy	11.30 – 15.00	1.02	0.87	1.05
97	bright	10.30 – 15.30	1.14	0.92	1.00
116	bright	11.00 – 13.30	0.57	0.59	0.48

Relative stomatal conductance during transient water stress

Because the absolute value of stomatal conductance is affected by many factors, relative stomatal conductance (RSC) is used as a measure to evaluate the effects of transient water stress. This parameter is obtained by expressing the absolute SC of treated leaves in per cent of the values of their controls (Fig. 1).

Initially after draining of the water, plant functioning was not affected because there was still enough water in the soil to meet the evaporative demands. The weather was fairly dull in the period between 56 and 74 DAP, but conditions improved from 74 DAP onwards. This improvement presumably coincided with the point where water became limiting. These circumstances caused a rapid fall in RSC. RSC values remained low during the stess period, the fluctuations in RSC between about 4 and 20% being associated with fluctuations in daily evaporative demand.

Upon rewatering in the evening on 90 DAP, RSC increased rapidly. After 96 DAP, RSC values generally exceeded 100%.

It is important to note that there were no clear differences between varieties in the pattern of decline of RSC during drying and in pattern of recovery immediately upon relief of stress.

The relation between leaf water potential and relative water content

Fig. 2 shows the relation between leaf water potential, PSI, and relative water content, RWC. Each data point represents a single leaf. The measurements were made on several days in controls and in stressed plants (during stress and after its relief). All data points conform to the same pattern, indicating that the relation between PSI and RWC was not affected by variety or transient stress.

The relation between osmotic potential and water potential

Fig. 3 shows the relation between leaf water potential, PSI, and osmotic potential, PSI. Each data point represents a single leaf. As in Fig. 2, data from different dates of measurement are not represented separately in Fig. 3, because all data points conformed to a single

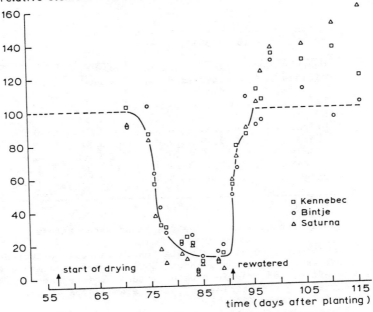

Fig. 1. Change with time of relative stomatal conductance during transient water stress.

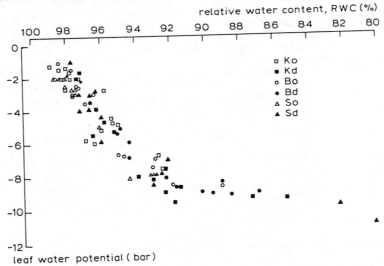

Fig. 2. Relation between leaf water potential and relative water content. The data were collected on 77, 89 and 92 days after planting. K = Kennebec, B = Bintje, S = Saturna, o = control (optimal water), d = transient drought treatments.

pattern. Indeed, the relation between PSI and PI did not differ between varieties and was not affected by transient drought.

The data points of controls with PSI values lower than -10 bar were obtained by artificial drying of the leaves in the pressure chamber. These points fall on the same curve as points from naturally stressed leaves. This supports the conclusion that the drought treatment did not exert specific effects on the osmotic behaviour of the varieties tested.

During stress (88 DAP) an additional test on osmotic adjustment was made. Six leaves of each treatment were excised and saturated to regain full turgor. After a saturation period of three hours, osmotic potential was determined (Table 2). PI values of saturated leaves did not differ between varieties and treatments. These results provide additional evidence for the absence of specific effects of water stress on osmotic behaviour.

Because the relations between PSI and RWC, and between PI and PSI were similar, the other relations of the Höfler-Thoday diagram, namely P-RWC and PI-RWC, will not be affected by variety or treatment either (data not presented).

Table 2. Osmotic potentials (bar) of control and stressed leaves after saturation with water for three hours[a]. Data were collected on 88 days after planting (n = 6).

	Kennebec		Bintje		Saturna	
	control	stressed	control	stressed	control	stressed
Mean:	-7.5	-7.7	-7.9	-7.8	-7.7	-8.2
S.E.:	0.1	0.2	0.2	0.2	0.1	0.1

a leaf water potentials at the time of excision were about -4 to -6 bar in control leaves and between -8 and -10 bar in stressed leaves.

Discussion

The main conclusions from the present study are:
1. Leaf water potential of cvs Kennebec, Bintje and Saturna are similar when ample soil water is available.
2. With optimal supply of water, stomatal conductance was somewhat lower in Bintje than in Kennbec and Saturna, except for the last weeks when Saturna appeared to senesce faster. Whether these differences in SC resulted in differences in water use efficiency remains to be analyzed (but: see Bodlaender, this volume).
3. Transient stress did not affect the leaf water relations in any of the varieties tested.
4. There were no indications of osmotic adjustment during stress.

The present results contrast with those reported by Levy (1983). This author observed higher water potentials and osmotic potentials, and lower stomatal conductance in cv. Up-to-Date than in cv. Désirée, both under non-stress and stress conditions. Furthermore, Levy concluded that there were differences among varieties in the degree of osmotic adjustment in response to stress.

The differences between our conclusions and those of Levy may arise from the fact that we examined different varieties. Differences in experimental conditions can also play a role.

Our approach differed from Levy's at two points. Levy made inferences on the degree of osmotic adjustment from calculations of the ratio of the

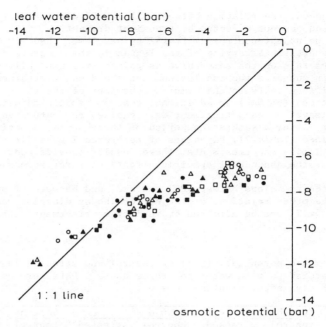

Fig. 3. The relationship between osmotic potential and leaf water
potential. The data were collected on 77, 89, and 92 days after
planting. See Fig. 2 for the explanation of symbols.

arithmetic differences of PI and PSI between stressed and unstressed
plants. The changes in PI and PSI were relatively small, namely maximally
1.65 and 2 bar, respectively. Considering the errors involved, our
approach to examine the occurrence of osmotic adjustment is presumably
more reliable, because we used measurements of changes in PSI, PI and P
(pressure potential, or turgor) over the entire range of values between
full turgidity till beyond turgor loss, moreover we included determinations of the relative water content. The second point concerns the
measurement of stomatal conductance. We used a steady state porometer,
whereas Levy used a porometer that measures the resistance to diffusion
of krypton across a leaf. The value of this trans-leaf diffusion resistance is affected by the frequency of stomates on the upper and the lower
leaf surface. Dwelle et al. (1983) found these frequencies to differ
between cultivars. Furthermore, the differences in stomatal conductance
in unstressed Up-to-Date and Désirée plants was about a factor two. This
difference seems too large to be physiologically realistic. Perhaps
Up-to-Date is a variety with extremely few stomata on the upper leaf
surface.

   Still, the remark about Levy's method to determine stomatal conductance
is not meant to deny varietal diffences in this parameter, because such
differences were indicated in the present study and by Dwelle et al.
(1981, 1983). However, the significance of such differences is a matter
of further study. It is also relevant to note that Shimshi & Susnoschi
(1985) reported conspicuously high values of leaf water potential under
water stress for cv. Up-to-Date in comparison to other cultivars.

   A conclusion of this discussion can be that further study is needed on
the methods to determine plant water status parameters and stomatal
conductance, on the effects of experimental conditions on the parameters,

34

and on the significance of varietal differences in parameter values with
respect to drought tolerance.

References

Commissie voor de Samenstelling van de Rassenlijst voor Landbouwgewassen,
1985. 60e Beschrijvende Rassenlijst voor Landbouwgewassen.
Leiter-Nypels, Maastricht.

Dwelle, R.B., G.E. Kleinkopf, R.K. Steinhorst, J.J. Pavek & P.J. Hurley,
1981. The influence of physiological processes on tuber yield of potato
clones (Solanum tuberosum L.): stomatal diffusive resistance, stomatal
conductance, gross photosynthetic rate, leaf canopy, tissue nutrient
levels, and tuber enzyme activities. Potato Res. 24: 33-47.

Dwelle, R.B., P.J. Hurley & J.J. Pavek, 1983. Photosynthesis and stomatal
conductance of potato clones (Solanum tuberosum L.). Comparative
differences in diurnal patterns, response to light levels, and
assimilation through upper and lower leaf surfaces. Plant Physiol. 72:
172-176.

Levy, D., 1983. Varietal differences in the response of potatoes to
repeated short periods of water stress in hot climates. 1. Turgor
maintenance and stomatal behaviour. Potato Res. 26: 303-313.

Morgan, J.M., 1980. Differences in adaptation to water stress within crop
species. In: N.C. Turner & P.J. Kramer (Eds): Adaptation of plants to
water and high temperature stress. John Wiley & Sons, New York.
p. 369-382.

Shimshi, D. & M. Susnoschi, 1985. Growth and yield studies of potato
development in a semi-arid region. 3. Effect of water stress and
amounts of nitrogen top dressing on physiological indices and on tuber
yield and quality of several cultivars. Potato Res. 28: 177-191.

Turner, N.C. & M.M. Jones, 1980. Turgor maintenance by osmotic
adjustment: a review and evaluation. In: N.C. Turner & P.J. Kramer
(Eds): Adaptation of plants to water high temperature stress. John
Wiley & Sons, New York. p. 87-103.

EFFECTS OF DROUGHT ON WATER USE, PHOTOSYNTHESIS AND TRANSPIRATION OF
POTATOES 1. DROUGHT RESISTANCE AND WATER USE

K.B.A. Bodlaender

Centre for Agrobiological Research (CABO), Wageningen, The Netherlands

Summary

   The relationship between drought resistance of eight potato varieties
(according to the Dutch list of varieties) and their rate of water use
was studied in a glasshouse experiment. Such a relation could not be
demonstrated, however, with increasing drought resistance dry matter
yields increased and water use efficiency improved. Indications were seen
for a relation between total water use and dry matter production.

Keywords: drought resistance, dry matter production, water use
efficiency.

Introduction

   Distinct differences in response to periods of drought are observed
between potato varieties. For instance, the figure for drought resistance
(or tolerance) in the Dutch list of varieties (Rassenlijst 1980) varies
from 5 to 9; this figure is mainly determined by the reaction of the
foliage of potato crops to drought. Varietal differences in response to
drought were also observed by several authors (Roztropowicz, 1978;
Steckel & Gray, 1979; Levy, 1983a and b and Susnoschi & Shimshi, 1985).
The effect of drought on tuber yield was mainly presented; Steckel & Gray
(1979) showed also the effect on total dry weight. The influence of
drought on tuber yield is caused by the effects of drought on total dry
matter production and in some cases also by the effect on distribution of
dry matter to foliage and tubers.
   Drought resistance of a variety can be brought about in several ways
(e.g. a low rate of water use by closure of stomata, drought avoidance by
a large root system or drought tolerance by osmotic adjustment and rapid
recovery after a drought period (van Loon, 1981; van Loon, 1985).
   These different responses to drought may be of importance under
different conditions: in regions where in general only short periods of
drought occur (as in the Netherlands), drought avoidance by a large root
system and rapid recovery after a drought period are most favourable to
growth and production; in regions where long periods of drought occur
(for instance in some tropical countries) drought resistance caused by a
low rate of water use may be most effective.
   To study the relationship between drought resistance and the rate of
water use - before, during and after a period of drought - a glasshouse
experiment with eight potato varieties was carried out. The relations
between drought resistance and total dry matter production and between
total dry matter production and total water use were also studied.

Materials and methods

   Plants were grown in a glasshouse at nearly constant temperature
(15-17 °C). Daylength was extended with electric bulbs during the first
weeks of growth.

The experiment was carried out with eight varieties with divergent drought resistance (according to the Dutch list of varieties): Saturna, Provita, Alpha, Astarte, Cardinal, Prevalent, Arran Banner and Bintje.

Seed tubers of these varieties were planted on March 19, 1980 in bins (size 90 cm x 70 cm x 40 cm). The bins were filled with "white" sand; at the bottom of the bins a water layer could be maintained. Per variety three bins were planted with each six seed tubers. For the measurements of water use three replicates were used; two of these were also used for the determination of total dry weight and tuber yield and the third replicate for additional observations.

Evaporation from the soil was prevented by covering the sand in the bins with a plastic sheet. The rate of water use was determined by weighing the bins (with plants) once or twice a week. The amount of water transpired per bin was supplied to the bins each time.

The plants were regularly supplied with water till April 25. Then, a period of drought was started: the water at the bottom of the bins was drained away. The transpiration of the plants was determined also during the drought period. On May 21 water was again supplied to the bins. At that moment the water content in the soil was very low (c. 1%) and all leaves of the plants were heavily wilting, except those of variety Prevalent. After rewatering the plants recovered from the drought within 1 or 2 days.

On June 16 the plants were harvested; on that day the foliage was in general still green. Fresh weights and dry matter contents of leaves, stems, tubers and other underground parts were determined. The leaves of variety Arran Banner were diseased in the later part of the experiment; results of this variety are presented, but generally not discussed.

The opening of the stomata was determined with the infiltration-technique (described by van Loon & Glas, 1978) shortly before, during and a few days after the drought period, always at about 15.00 p.m., when leaf water potentials were also determined with the pressure-bomb.

Results

About 50 days after planting maximal plant length was reached. Distinct differences in plant length were observed between the varieties: Astarte had the tallest plants (c. 120 cm); on the other hand the stems of Saturna, Alpha and especially Prevalent remained rather short (c. 60 cm).

The water use per day is presented in Figure 1. The differences in water use during the first weeks may be attributed to differences in rate of development between the varieties. During the drought period the varieties Saturna, Alpha and Astarte were the first to show signs of wilting; this will be due to their high water use. After the 50st day of the growing period water use of all varieties decreased rapidly, because the water in the sand was exhausted. The water use of Prevalent was less decreased: this variety had generally a lower water use and therefore the water content of the soil remained higher during the dry period. After rewatering the water use of all varieties (except of the diseased Arran Banner) increased rapidly in a few days; for some varieties, however, water use after rewatering remained somewhat lower than before the drought period.

The stomatal opening (Fig. 2) of all varieties decreased distinctly during the drought period; the stomata of variety Alpha closed most rapidly, those of Prevalent most slowly. After rewatering the stomata of all varieties reopened rapidly.

The leaf water potentials of all varieties (Fig. 3) became much lower during the drought period; highly negative values were obtained. Saturna

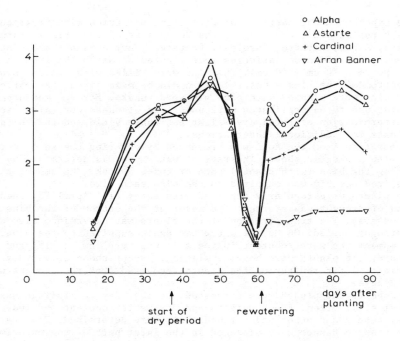

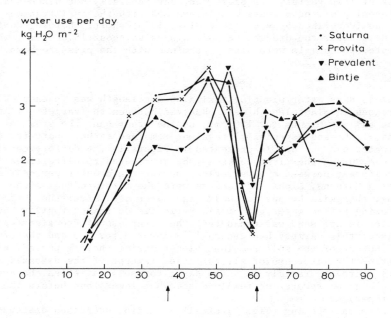

Fig. 1. Water use per day of 8 potato varieties before, during and after
a dry period.

38

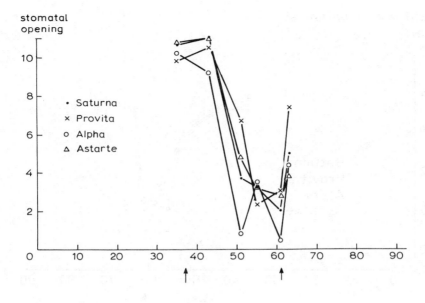

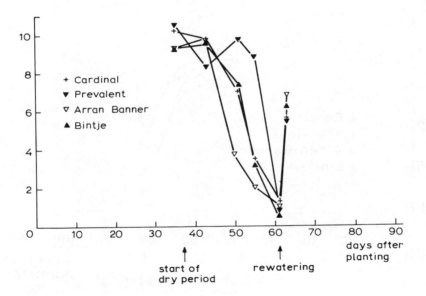

Fig. 2. Stomatal opening (determined at about 15.00 p.m.) of 8 potato
varieties shortly before, during and after a dry period.

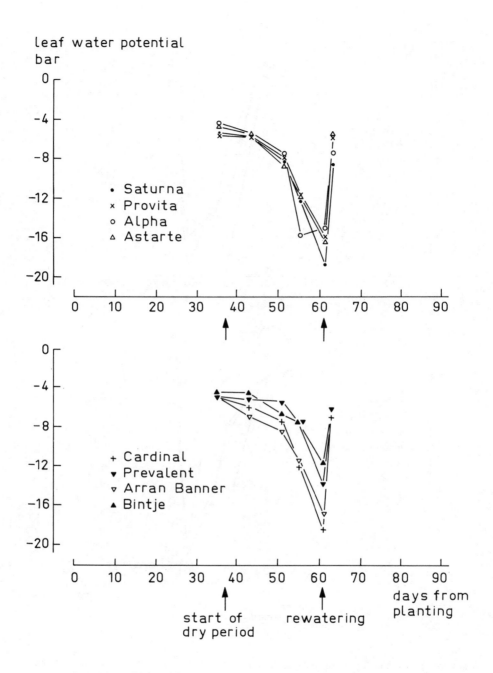

Fig. 3. Leaf water potential (determined at about 15.00 p.m.) of 8 potato
varieties shortly before, during and after a dry period.

and Cardinal showed the most negative leaf water potential and those of Prevalent and Bintje were the least negative. After rewatering the leaf water potentials returned in general to values comparable with those of before the drought period.

The relations between drought resistance of the eight varieties according to the Dutch list of varieties and their total water use and total dry matter production in this experiment are shown in Table 1. Obviously, no clear relation exists between total water use in this experiment and drought resistance of these varieties; the results are rather variable (compare for instance Alpha and Bintje). In this experiment indications were seen for an increase in total dry weight with increased water use; however, large differences in total dry weight were found between varieties with nearly the same total water use (compare Saturna and Bintje).

The relationship between drought resistance of the varieties and their water use efficiency in this experiment is shown in Figure 4. The water use efficiency (= kg water used for the production of 1 kg dry matter) was highly correlated with the drought resistance of the varieties (the diseased Arran Banner excepted).

Tuber dry weights of the varieties studied showed in general the same order as the total dry weights (Table 1). However, the relation between drought resistance and water use efficiency for tuber dry weight (Fig. 4), showed an unfavourable deviation for variety Astarte, in other words Astarte had a more unfavourable dry matter distribution than the other varieties.

Table 1. Relation between drought resistance, water use, dry matter production and tuber yield of 8 potato varieties.

Varieties	Drought[1] resistance	Total[2] water use kg/m^2	Total[3] dry weight kg/m^2	Tuber[3] dry weight kg/m^2
Saturna	5	195	1.05	0.80
Provita	6	179	1.17	0.87
Alpha	7	217	1.53	1.14
Astarte	7	209	1.35	0.83
Cardinal	8	187	1.32	1.01
Prevalent	8	172	1.22	0.86
Arran Banner*	9	139	0.73	0.47
Bintje	9	199	1.58	1.24

1) See Dutch list of varieties, 10 = resistant
2) During 3 months
3) 3 months after planting
* Later diseased

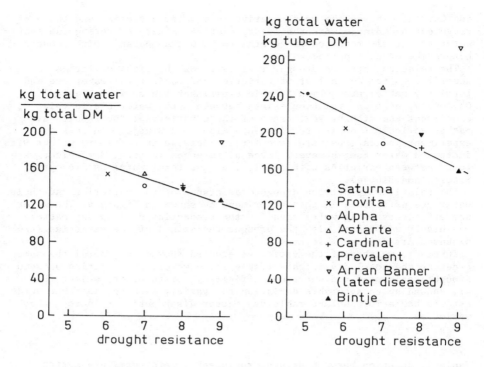

Fig. 4. The relationship between drought resistance of 8 potato varieties and
left: water use efficiency (= kg total water used per kg total dry matter produced during 3 months including a drought period); r = -0.92 (without Arran Banner)
right: water use efficiency for tuber dry weight (= kg total water used per kg tuber dry matter produced during 3 months including a drought period); r = -0.74 (without Arran Banner).

Discussion

A relation between drought resistance of potato varieties and their water use efficiency was observed; the relationships between drought resistance and total water use or total dry matter production were less distinct or absent. However, these values for total dry matter yield and total water use were obtained after three growth periods: normal water supply, increasing drought and recovery from drought after rewatering.
Steckel & Gray (1979) found in the dry treatments of their field experiments some relation between total yield of dry matter and cumulative water use, however, the differences between varieties were not clear. The amounts of supplied water and the experimental year had distinctly larger effects than the differences between varieties. Rijtema & Endrödi (1970) found in irrigation experiments in several years with different varieties no distinct relationship between total dry matter yield and total transpiration. Weather and soil conditions in the different experiments undoubtedly influenced the results.

It is evident that further information is needed about the relationship between drought resistance of potato varieties and their water use and dry matter production under various conditions. These relationships should be studied during different growth phases and during periods of drought and of sufficient water supply. The relation between drought resistance of varieties and their water use efficiency may give interesting information about the response of these varieties to drought.

Acknowledgement

Thanks are due to P.A. te Velde for carrying out the glasshouse experiment.

References

Levy, D., 1983a. Varietal differences in the response of potatoes to repeated short periods of water stress in hot climates. 1. Turgor maintenance and stomatal behaviour. Potato Research 26: 303-313.

Levy, D., 1983b. Varietal differences in the response of potatoes to repeated short periods of water stress in hot climates. 2. Tuber yield and dry matter accumulation and other tuber properties. Potato Research 26: 315-321.

Loon, C.D. van, 1981. The effect of water stress on potato growth, development and yield. American Potato Journal 58: 51-69.

Loon, C.D. van, 1985. Drought, a major constraint in potato production. In: "Potato Research of Tomorrow" focusing on: virus resistance, drought tolerance, analytic breeding methods. International Seminar 30 and 31 October 1985. Pudoc, Wageningen, the Netherlands: this publication.

Loon, C.D. van & Tj.K. Glas, 1978. Comparison of some methods to characterize the water status of the potato plant in the field. EAPR Abstracts of Conference Papers. 7th Triennial Conference of the European Association for Potato Research. Warsaw, Poland: 34-35.

Rassenlijst 1980. 55e Beschrijvende Rassenlijst voor Landbouwgewassen. Wageningen: 1-336.

Roztropowicz, S., 1978. Some aspects of Polish physiological and agrotechnical research on the potato. Survey papers 7th Triennial Conference EAPR: 35-60.

Rijtema, P.E. & Endrödi, 1970. Calculation of production of potatoes. Netherlands Journal of agricultural Science 18: 26-36.

Steckel, J.R.A. & D. Gray, 1979. Drought tolerance in potatoes. Journal of agricultural Science, Cambridge 92: 375-381.

Susnoschi, M. & D. Shimshi, 1985. Growth and yield studies of potato development in a semi-arid region. 2. Effect of water stress and amounts of nitrogen top dressing on growth of several cultivars. Potato Research 28: 161-176.

EFFECTS OF DROUGHT ON WATER USE, PHOTOSYNTHESIS AND TRANSPIRATION OF POTATOES. 2. DROUGHT, PHOTOSYNTHESIS AND TRANSPIRATION

K.B.A. Bodlaender, M. van de Waart and J. Marinus

Centre for Agrobiological Research (CABO), Wageningen, The Netherlands

Summary

Irrigation does not always influence potato production positively. Therefore the effect of drought on dry matter production and water use was studied by measuring photosynthesis and transpiration of potato crops comparing dry and wet plots (net area 2 m^2). Water shortage diminished both net photosynthesis and transpiration; transpiration, however, was more sensitive to drought than photosynthesis. No clear relationship was seen between water status of the plants (leaf water potential, stomatal opening) and the rate of net photosynthesis; other factors as rate of irradiance, temperature, air humidity may influence these parameters.

Keywords: drought, dry matter production, photosynthesis, transpiration, water status.

Introduction

The influence of water supply on tuber yield of potato crops was studied by many authors (see surveys Burton, 1966; Harris, 1978; van Loon, 1981). Often a relation was found between total amount of rainfall or irrigation water and tuber yield (especially dry weights), but in some cases no distinct relationship was seen.

The decrease in tuber yield by drought will be caused to a large extent by the effect of drought on dry matter production. Bodlaender et al. (1982) found an increase in total dry weight by extra water supply through irrigation, this increase being small in 1972 and large in 1973. Steckel & Gray (1979) also observed an increase in total yield of dry matter with increasing amounts of cumulative water used. Versteeg (1985) found in irrigation experiments in Peru a distinct increase in total dry weight with increasing amounts of irrigation. Penman (1963) described a direct relation between tuber yield and adjusted potential transpiration with different varieties and amounts of irrigation.

The effects of drought on total dry matter production will be caused by the effect of water shortage on photosynthesis per unit land area. Extra water supply by irrigation or infiltration showed variable results. The question arises under which moisture conditions irrigation affects total dry matter production and tuber yield positively. Therefore attempts were made to determine the water status of the plants at which a distinct reduction in photosynthesis will occur: if suitable parameters for the early detection of stress can be found, irrigation can be supplied timely. The effect of water supply or drought on net photosynthesis (dry matter production) and transpiration (water use) was studied in some experiments in which potato plants were grown in semi-crop conditions; therefore the results may be extrapolated to normal crop growth under field conditions.

Figure 1. Measurement of photosynthesis and transpiration of a dry and a
wet plot in transparent enclosures. Plants were grown under the
transparent roof in the background till the start of the
measurements. Experiment 1982.

Materials and methods

In 1978, 1979 and 1982 photosynthesis and transpiration were measured
of potato plots with sufficient water supply (wet plots) and with drought
increasing during some days (dry plots). These measurements were carried
out in several weeks on different plots with different starting dates of
the dry periods. Vos (1985) continued these investigations after 1982
studying also the recovery after a drought period.

In the first two experiments pits of 18 m^2 were dug to a depth of 60 cm
(1978) or 50 cm (1979). The bottom and the walls of the pits were covered
with plastic sheets to ensure hydraulic isolation of the pits from the
surrounding soil. Light expanded clay granules were laid on these sheets
at the bottom of the pits to obtain a water layer. The pits were refilled
with humous sand. All plots were covered by a transparent roof to shelter
the plants from rain. Water was supplied to the plants from the water
layer at the bottom of the plots; this layer was regularly replenished.
The dry plots, however, were not supplied with water for some weeks
before and during the measurements. The plants could grow through nylon
nets (mesh 10 cm x 10 cm) at 50 cm height to prevent heavy lodging.

In 1982 the plants were grown in bins of 2 m^2 and 40 cm depth. These
bins were filled with humous sand and also had a water layer at the
bottom. These bins were placed under a transparent roof. Several weeks
after planting – at successive times – bins were transported to the open
air for the measurements. Each plot consisted of 9 bins of 2 m^2. The
middle bin was used for the measurements; the plants in the other bins

were used as guard plants (see Fig. 1). Net photosynthesis (and respiration during the night) and transpiration of a wet and a dry plot were measured simultaneously in transparent plant chambers of 2 m^2. These plant chambers were placed on metal frames in the middle of the plots (1978 and 1979) or on the metal middle bin (1982). Photosynthesis and transpiration were measured continuously on the same plots during 2–3 days (1978 and 1979) or 4–5 days (1982); afterwards measurements were started on two other plots. After each period of measurements the plants were harvested. Leaf area and weights of the different parts of the plants were determined. The weights will not be discussed here; in general no distinct effects of drought on total dry weight or tuber yield were visible, the periods of drought were relatively short.

The measurements of photosynthesis and transpiration were carried out with mobile equipment, provided with infra-red gas analysers, to determine the CO_2 concentration and the humidity of the increasing and outgoing air of the plant chambers. The equipment and methods were described in detail by Louwerse and Eikhoudt (1975).

Some additional observations were made on the water status of soil and plants. The moisture tension of the soil was measured with tensiometers at 15 and 25 cm depth; in the first years water content of the soil was also determined at various depths. Several times leaf water potentials were determined with the pressure-bomb and the opening of the stomata with the infiltration technique (van Loon & Glas, 1978). Leaves for these determinations were taken in 1978 and 1979 from the guard plants outside the plant chamber and in 1982 from plants in the transparent enclosure.

In 1978 varieties Bintje and Saturna were used, in the other two years only variety Saturna. The number of plants per m^2 were 6 (Saturna 1978 and 1982) or 5 (Bintje 1978, Saturna 1979); so, each net plot consisted of 12 or 10 plants. Seed tubers were generally planted in mid-April (except the second planting at the end of May 1979).

Results

In this paper only some results will be presented of variety Saturna in the experiments of 1978 and 1982; the measurements with Bintje in 1978 and with Saturna in August 1979 showed less pronounced effects of drought.

Experiment 1978

The measurements of photosynthesis and transpiration started in mid-June; the soil was then completely covered by foliage. The water content in the soil was lower in the dry than in the wet plots, especially in the deeper soil layers; on June 19, for instance, the difference was 6–12%. This lower moisture content caused on the bright day June 20 some reduction in net photosynthesis in the dry plots; transpiration, however, was much more reduced by drought (Fig. 2). During that day irradiance increased to 800 W m^{-2} and the temperature to 25 °C; temperature in the enclosures followed the outside air temperature. Net photosynthesis at the same level of irradiance was lower in the afternoon than in the morning; this phenomenon was also observed in other experiments. Transpiration of the wet plot on June 20 did not show a hysteresis curve, as was often found on other days (see also Fig. 5). Both net photosynthesis and transpiration increased with increasing rates of irradiance; however, at higher levels of irradiance (500 W m^{-2} and higher) photosynthesis did not further increase whereas transpiration continued to increase. The leaf area index of the dry and the wet plots did not differ much.

46

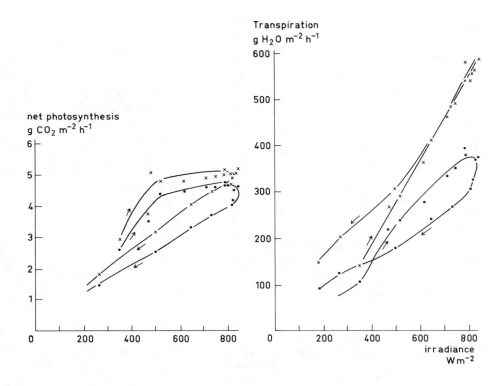

Figure 2. Light-response curves of net photosynthesis of a dry (.) and a
wet (x) plot on 20 June 1978. Variety Saturna. Arrows indicate
course of time.

Leaf water potentials (Fig. 3) decreased during the day from -2 to -7
bar and increased later again. The water status of the leaves was
affected by drought: the dry plot had slightly lower leaf water
potentials than the wet plot. The stomatal opening (Fig. 3) increased in
both plots till 10 a.m. Thereafter the stomatal opening of the dry plot
decreased slightly till 12 a.m. and later rapidly. The decrease in
stomatal opening in the wet plot started later (about noon). The
difference between both plots was rather large. The decreases in stomatal
opening, in transpiration and in net photosynthesis did not concur in
time and were not of the same order (Table 1). The decrease in net photosynthesis on the dry plot between 14.00 and 16.00 hours was still
relatively small with almost closed stomata.

Photosynthesis and transpiration were also measured at a later growth
stage (4-15 August). Net photosynthesis was more reduced by drought than
in June; this was partly due to the lower leaf area index in the dry
than in the wet plots. Transpiration was also distinctly decreased by
drought. The levels of net photosynthesis and transpiration were in
general distinctly lower in this older crop than in the younger one in
June.

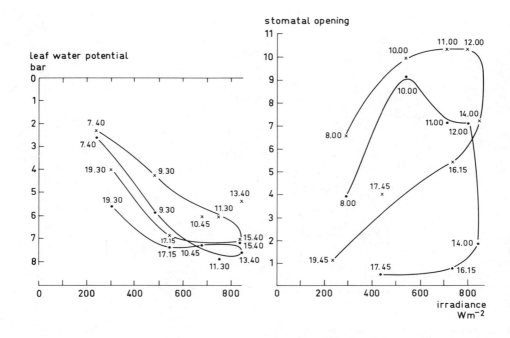

Figure 3. Light response curves of leaf water potential and stomatal
         opening (11 = stomata completely open) of a dry (.) and a wet
         (x) plot on 20 June 1978. Variety Saturna.

Table 1. Temperature and irradiance; stomatal opening, transpiration and
        net photosynthesis of a dry and a wet plot at 3 times on 20 June
        1978. Variety Saturna.

Time hour	Temp. °C	Irradiance $W\ m^{-2}$	Stomatal opening*		Transpiration $g\ H_2O\ m^{-2}\ h^{-1}$		Net photosynthesis $g\ CO_2\ m^{-2}\ h^{-1}$	
			D**	W**	D	W	D	W
c. 12.00	21.5	800	7.5	10.5	380	500	4.7	5.0
c. 14.00	23.5	840	2	7.5	380	550	4.6	5.0
c. 16.00	25.0	765	1	5.5	280	480	3.8	4.6

* 11 = stomata completely open
** D = dry   W = wet

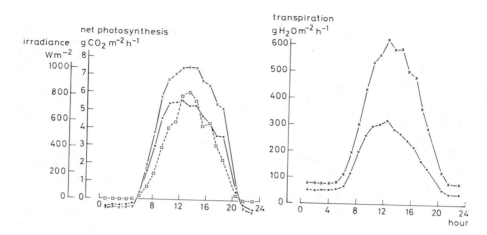

Figure 4. Net photosynthesis and transpiration of a dry (.) and a wet (x)
plot and irradiance (□) on 7 July 1982. Variety Saturna.

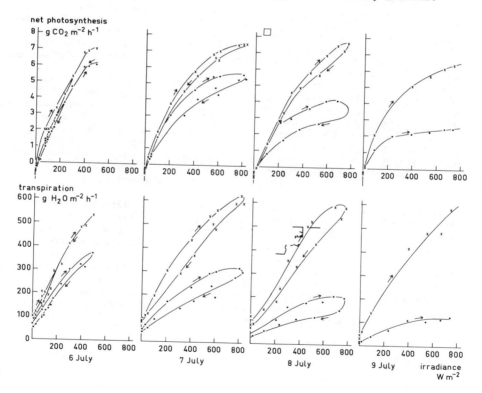

Figure 5. Light-response curves of net photosynthesis and transpiration
of a dry (.) and a wet (x) plot on 6-9 July 1982. Variety
Saturna. Arrows indicate course of time.

Experiment 1982

Photosynthesis and transpiration were measured during several weeks in
June and July; the foliage covered the soil completely. Because the soil
layer in the bins was only 35 cm deep and the water at the bottom of the
bins of the dry plots was drained away at the start of the dry periods,
the plants of the dry plots showed the first signs of drought within a
week. Every week other net plots were used for the measurements; the
drought periods of the different dry plots (bins) started at successive
times. The effect of drought increased during each period; at the end of
each week - especially in the later ones - wilting of the leaves became
obvious. Soil water potential at the start of the measurements of
photosynthesis and transpiration always was distinctly more negative in
the dry than in the wet plots and showed in the dry plots very high
values at the end of each week. The leaf area of the dry and the wet
plots did not differ much (except on July 9).

The temperature in the plant chambers during the measurements was in
general 20 °C. The responses of net photosynthesis and transpiration
during July 7 (a nice clear day) are presented in Figure 4. Net photosynthesis is distinctly reduced by drought; the effect on transpiration
is however, still greater. These effects can also be demonstrated by
plotting net photosynthesis and transpiration against irradiance (Fig.
5). These measurements were carried out on the same two plots from 5
(afternoon) till 9 (noon) July. The curves show that also in this week
both photosynthesis and transpiration were higher in the morning than in
the afternoon at the same level of irradiance. The data on the different
days show that the difference in net photosynthesis between the wet and
the dry plot increased distinctly during a week, and was substantial at
the end of the week when wilting of the leaves of the dry plot was
obvious. The reduction in transpiration by drought was also markedly
increased during this week. Comparable effects on net photosynthesis and
transpiration by drought were found during the periods 21-25 June and
28 June - 2 July; the effects during these weeks were somewhat less
pronounced than in the period 5-9 July. The harvest on July 9 showed a
considerably lower leaf area in the dry plot than in the wet one, caused
by the heavy drought in that period. However, the leaf area of the dry
and the wet plot did not differ distinctly at the end of the former two
periods.

Leaf water potentials and stomatal openings were determined at a
restricted number of times only. Leaf water potential (Fig. 6) was in
general lower on the dry than on the wet plot, in the week of 5-9 July,
especially so in the morning hours. Leaf water potentials of both plots
decreased rapidly till noon. The stomatal opening (Fig. 6) showed a
distinct difference between the dry and the wet plot; the stomata of the
dry plot were often closed to a large extent. In the week 28 June -
2 July a much greater difference in leaf water potential between the dry
and the wet plot was observed than in the week thereafter; a large
difference in stomatal opening between the two plots was seen in both
weeks.

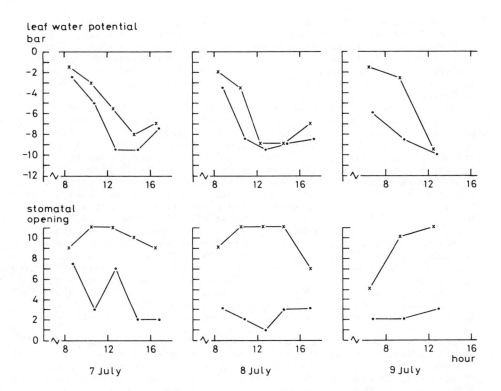

Figure 6. Leaf water and stomatal opening (11 = stomata completely open) of a dry (.) and a wet (x) plot on 7, 8 and 9 July 1982. See further Fig. 5.

Discussion

In the above described experiments an effect of drought on net photosynthesis could be observed, even when the plants did not yet show distinct signs of wilting. The reduction in photosynthesis increased with severity of the drought during successive days, and became substantial when many of the leaves were wilting. Drought reduced transpiration more than net photosynthesis. The response of photosynthesis diminished with increasing rates of irradiance during the morning, transpiration, however, did not show such a response. This result was in contrast to the statement of Pettite and Goltz (1982) that a decrease in net photosynthesis was linearly related to a decrease in transpiration under stress.

A reduction in photosynthesis by water stress - as seen in the above described experiments - was not found by Sale (1974); he observed in that case great differences in leaf water potential. A reduction was, however, observed by several other authors: Munns & Pearson (1974), Moorby et al. (1975), Ackerson et al. (1977), Ringel (1981), Pettite & Goltz (1982), Shimshi et al. (1983) and Shimshi & Susnoschi (1985); Shekhar & Iritani (1979) also found indications in that direction. Photosynthesis of several varieties was affected by drought in a different way (Wilcox & Ashley, 1982: 2 of 4 varieties showed a decrease by drought) or to a

different degree (Shimshi & Susnoschi, 1985). In the experiment of
Shimshi & Susnoschi water shortage showed a greater effect on
photosynthesis of leaves from 75 days old plants than on that of leaves
from 54 days old plants. Ackerson et al. (1977) observed under stress a
greater effect on photosynthesis of the lower leaves than on that of the
upper leaves; Moorby et al. (1975) found a greater effect of drought on
young leaves than on older ones.

The differences in relation to drought will not only be influenced by
variety and age of plants or leaves, but also by the methods used and the
growing conditions. Munns & Pearson (1974), Moorby et al. (1975),
Ackerson et al. (1977), Shekhar & Iritani (1979), Shimshi et al. (1983)
and Shimshi & Susnoschi (1985) used the $^{14}CO_2$ method with separate
leaves; Wilcox & Ashley (1982) determined photosynthesis with leaf disks
in a Warburg apparatus. Only Sale (1974) determined the CO_2 uptake of a
canopy in a transparent plant chamber with an infra-red gas analyser as
in our experiments. In some experiments plants were grown in pots in
greenhouses or growth chambers (Moorby et al., 1975; Shekhar & Iritani,
1979; Wilcox & Ashley, 1982), in other cases the effect of drought on
photosynthesis was studied in field experiments with different irrigation
treatments (Sale, 1974; Ackerson et al., 1977; Shimshi & Susnoschi,
1985). In some cases different moisture conditions were maintained during
long periods (for instance Shimshi & Susnoschi, 1985) or short periods of
5-15 days (for instance Moorby et al., 1975; Wilcox & Ashley, 1982). This
last situation occurred in our experiments, especially in 1982. Therefore
adaptation of the plants to drought will not have played a role of any
importance.

In the experiments of 1978 and 1982 the influence of drought on the
status of the leaves was observed. Drought decreased leaf water potential
somewhat, the effect on stomatal opening was more pronounced. No clear
relationship between leaf water potential and net photosynthesis could be
demonstrated in these experiments. Partial closure of stomata was
observed before a reduction in net photosynthesis started.

According to Shimshi et al. (1983) leaf permeability (determined with a
porometer) seems a better index than leaf water potential for
characterization of water stress. Ackerson et al. (1977) demonstrated a
distinct relationship between leaf diffusive resistance and leaf water
potential but the data showed a large scatter. Other authors (see Vos,
1985; van Loon, 1983) also consider relative stomatal resistance or
conductance (determined with a porometer) a good criterion for the water
status of plants. It seems, however difficult (or not yet possible) to
determine at which absolute water status of the leaves a substantial
decrease in net photosynthesis will occur and as a consequence extra
water supply will be needed. Water status of leaves is not only
influenced by soil moisture tension, but also by other factors as
irradiance, and vapour pressure deficit. Therefore leaf water status of a
dry and a wet plot should be determined at the same time under the same
conditions. In stead of the indirect detection via the water status of
leaves, a reduction in net photosynthesis of leaves can now easily be
determined with a portable leaf chamber analyser; also with this method
the above mentioned other factors will influence the measurements. A
direct comparison of the various methods to determine the effects of
drought on photosynthesis should be carried out.

Short periods of drought will not have a great effect on dry matter
production; however, later in the season - due to aging or dropping of
leaves - drought will affect production more negatively.

Acknowledgements

Thanks are due to mr W. Louwerse and mr J. van Kleef for carrying out the measurements of photosynthesis and transpiration with the mobile equipment and miss E.G. Veenhof for observations on water status of the plants in 1982.

References

Ackerson, R.C., D.R. Krieg, T.D. Miller & R.G. Stevens, 1977. Water relations and physiological activity of potatoes. Journal of the American Society of Horticultural Science 102: 572-575.

Bodlaender, K.B.A., J. Marinus & M. van der Waart, 1982. Het effect van stikstofbemestingen in verschillende ontwikkelingsstadia op groei en opbrengst van aardappelen op zand- en dalgrond. Stikstof 101: 9-16.

Burton, W.G., 1966. The potato. H. Veenman & Zonen NV, Wageningen, Holland: 66-81.

Harris, P.M., 1978. Water. In: P.M. Harris (Ed.): The potato crop. Chapman & Hall, London: 245-277.

Loon, C.D. van, 1981. The effect of water stress on potato growth, development and yield. American Potato Journal 58: 51-69.

Loon, C.D. van, 1985. Drought, a major constraint in potato production. In: Stichting voor Plantenveredeling SVP (Ed.): "Potato Research of Tomorrow" focusing on: Virus resistance, drought tolerance, analytic breeding methods. International Seminar 30 and 31 October 1985. Pudoc, Wageningen, the Netherlands: this publication.

Loon, C.D. van & Tj.K. Glas, 1978. Comparison of some methods to characterize the water status of the potato plant in the field. EAPR Abstract of Conference Papers. 7th Triennial Conference of the European Association for Potato Research. Warsaw, Poland: 34-35.

Louwerse, W. & J.W. Eikhoudt, 1975. A mobile laboratory for measuring photosynthesis, respiration and transpiration of field crops. Photosynthetica 9: 31-34.

Moorby, J., Munns, R. & J. Walcott, 1975. Effect of water deficit on photosynthesis and tuber metabolism in potatoes. Australian Journal of Plant Physiology 2: 323-333.

Munns, R. & C.J. Pearson, 1974. Effect of water deficit on translocation of carbohydrate in Solanum tuberosum. Australian Journal of Plant Physiology 1: 529-537.

Penman, H.L., 1963. Weather and water in the growth of potatoes. In: J.D. Ivins, F.L. Milthorpe (Ed.): The growth of the potatoes. Butterworths, London: 191-198.

Pettite, J.M. & S.M. Goltz, 1982. Effect of cyclic water stress on photosynthesis and transpiration of potato (Solanum tuberosum cv. Bel Rus). American Potato Journal 59: 482-483.

Ringel, B., 1981. Measurements of the daily course of photosynthesis and transpiration in potato leaves (Solanum tuberosum L.) under controlled conditions of light, temperature, CO_2 concentration and soil moisture. Archiv für Acker- und Pflanzenbau und Bodenkunde 25: 611-619.

Sale, P.J.M., 1974. Productivity of vegetable crops in a region of high solar input. III Carbon balance of potato crops. Australian Journal of Plant Physiology 1: 283-296.

Shekhar, V.C. & W.M. Iritani, 1979. Influence of moisture stress during growth on $^{14}CO_2$ fixation and translocation in Solanum tuberosum L. American Potato Journal 56: 307-311.

Shimshi, D., J. Shalhevet & T. Meir, 1983. Irrigation regime effects on some physiological responses of potato. Agronomy Journal 75: 262-267.

Shimshi, D. & M. Susnoschi, 1985. Growth and yield studies of potato development in a semi-arid region. 3. Effect of water stress and amounts of nitrogen top dressing on physiological indices and on tuber yield and quality of several cultivars. Potato Research 28: 177-191.

Steckel, J.R.A. & D. Gray, 1979. Drought tolerance in potatoes. Journal of agricultural Science 18: 28-36.

Versteeg, M.N., 1985. Factors influencing the productivity of irrigated crops in Southern Peru, in relation to prediction by simulation models. Thesis Wageningen, the Netherlands: 1-182.

Vos, J., 1985. Research on water relations and stomatal conductance in potatoes. 1. An introduction into concepts, techniques and procedures. In: Stichting voor Plantenveredeling SVP (Ed.). "Potato Research of Tomorrow" focusing on: virus resistance, drought tolerance, analytic breeding methods. International Seminar 30 and 31 October 1985. Pudoc, Wageningen, the Netherlands: this publication.

Wilcox, D.A. & R.A. Ashley, 1982. The potential use of plant physiological responses to water stress as an indication of varietal sensitivity to drought in four potato (Solanum tuberosum L.) varieties. American Potato Journal 59: 533-545.

A POSSIBLE SCREENING TECHNIQUE FOR DROUGHT TOLERANCE IN POTATO

A.G.B. Beekman and W.F. Bouma

Foundation for Agricultural Plant Breeding (SVP), Wageningen,
The Netherlands

Summary

Drought tolerance is considered to be a major component of yield
stability of a potato variety. The possibility of a large scale screening
technique for drought tolerance based on visual plant response to drought
stress was studied under glasshouse conditions. To this end potted potato
plants were subjected to drought stress, followed by rewatering. The
extent of wilting during the drought period and the potential for
recovery after rewatering were assessed. Large differences between
genotypes were recorded for wilting and recovery. For 47 varieties
studied, the correlation coefficients of the scores for wilting and
recovery with the scores for drought tolerance in the variety list were
$r = 0.57$ and $r = 0.64$ respectively.
It is concluded that the potential for recovery after a period of
severe drought stress under glasshouse conditions can be used to estimate
drought tolerance of a potato clone. For some genotypes the scores are
not consistent with field data, suggesting that the test does not include
all components of the drought tolerance complex. The results indicate the
existence of different mechanisms of drought tolerance. The possible use
in breeding of the greenhouse screening technique described is discussed.
Keywords: _Solanum tuberosum_, potato, drought tolerance, screening
technique, potato breeding, components of drought tolerance.

Introduction

Among the factors limiting potato yield, drought is considered of
major importance (Van Loon, 1981). Even in growing areas with a maritime
climate, the potato crop is often exposed to (short) periods of drought
stress. Therefore, drought tolerance is considered an important component
of yield stability. It should be noted that most potato varieties of wide
geographic distribution are drought tolerant.
Levitt (1972) and Jones et al. (1981) distinguished drought escape (or
avoidance) and drought tolerance per se. In this paper, however, this
distinction is not made and the term drought tolerance refers to the
potential of a variety to yield well under conditions of drought stress.
The yielding ability under dry conditions is considered to be based on a
combination of drought tolerance sensu stricto and drought avoidance as
well.
To improve drought tolerance of the potato by breeding, techniques for
large scale screening are essential. In the literature many techniques to
detect variation in drought tolerance are described. The methods for
measuring differences in drought tolerance can be classified into three
groups:
1. Field techniques
The yield of field grown plants under different water regimes is
measured. Differences in the water supply may be caused by natural
differences in rainfall or may be caused artificially. Reliable estimates
of drought tolerance can be obtained by these methods. A drawback of
tests under natural conditions is that they depend on weather conditions.

Therefore under unstable weather conditions field tests need to be
replicated in space and time and are consequently laborious and time
consuming. Hence field tests are less suitable for large scale screening,
for example for breeding purposes.
2. Methods measuring physiological, biochemical or morphological
   characteristics of plants
   The most important methods are described extensively by Vos and by Van
Loon (this volume). All these techniques have in common that separate
traits influencing drought tolerance are measured. The assessment of
components of tolerance to drought may enable the selection of parents
combining different mechanisms of drought tolerance.
3. The third group of techniques is a compromise between the two groups
of test techniques mentioned. In these tests plants are subjected to a
controlled drought stress and the response of plants is recorded by
visual assessment.
   In the period 1979 – 1980 a greenhouse test for appraising the level
of drought tolerance of potato was developed by Van der Wal (unpublished)
at the Foundation for Agricultural Plant Breeding (SVP). The main results
have been presented earlier by Miedema (1984). In this paper the test
method is described and data obtained by Van der Wal and Bouma are given.

Material and methods

   The response of potato plants to drought stress was studied under
greenhouse conditions. For that purpose experiments were carried out with
two different groups of potato clones. To gain information on the
reproducibility of the results each group of genotypes was investigated
in a series of three experiments.
   In the period 1979 – 1980 three experiments I, II an III, were
conducted by Van der Wal, using 52 potato varieties, 47 of them having a
known score for drought tolerance ranging from 5 (very sensitive) to 9
(very tolerant). A second series of experiments (IV, V and VI) was
performed in the period 1981 – 1983 by Bouma, using 46 progenitors of the
Foundation for Agricultural Plant Breeding with unknown drought
tolerance. Three varieties, viz. the very drought tolerant Katahdin, the
rather drought tolerant Mentor and the very drought sensitive variety
Veenster, were included as standards. The experiments were carried out
from March to April in a greenhosue (I, III, IV, V an VI), and in the
period September to October (II).
   Plants were grown in plastic pots, standing on capillary matting of
cotton. The temperature during the experiments fluctuated between 15 ^{o}C
and 30 ^{o}C, with extremes ranging from 10 ^{o}C to 40 ^{o}C. In each experiment
the potted plants were subjected to one of the following treatments:
1. control or "wet" treatment, with ample water supply
2. "dry" treatment, where the water supply was interrupted during a fixed
   period, thus causing severe drought stress.

   There were two fully randomized replications per treatment, in which
each variety was represented by a row of four plants. The three standard
varieties in the second series of experiments were represented by two
rows of plants in each replication.
   To reduce the effect of seed tubers, plants were grown from eye
pieces, i.e. eyes with a half-spherical piece of tuber tissue of about 7
g. Eye pieces with sprouts of 0.5 cm cut from tubers, presprouted for 14
days at 18 ^{o}C, were planted in plastic pots (15 cm diameter, contents 2.7
l) filled with a mixture of sandy loam and peat based potting compost.
Per plant all stems but one were removed, to increase uniformity.

Assessments were made on two plants, Nos 2 and 3, in a row: the first and the last plants in each row of four plants formed two border rows. If necessary, pots within each row were exchanged, so that the experiment was started with two representative middle plants. Per plant all stems but one were removed, to increase uniformity.

Forty days after planting watering was stopped in the "dry" treatment. Then most plants had eight leaves and were about to flower. Some varieties had started tuber formation. When plants of the variety Katahdin were wilted for 75 per cent (and those of Veenster were completely wilted), plants in the "dry" treatment were rewatered. Generally six days after stopping the water supply the fist symptoms of drought stress were visible and after another seven to twelve days plants of Katahdin and Veenster had reached the wilting stage described above. The following assessments were made on plants in both treatments:
- stage of development four days before, three days after the beginning and at the end of the dry period.
- wilting was recorded on a scale presented in Table 1 at several dates during the dry period.
- recovery was recorded on the same scale as that for wilting 24 hours after rewatering.
- at the end of the experiments plant weight, plant length and weight of tubers were assessed.

Table 1. Scale used to score plants for wilting symptoms and the potential for recovery after severe drought stress under greenhouse conditions.

Score for wilting and recovery	Estimation of percentage of leaf area still turgescent	Description of the symptoms
9	> 95	all leaves turgescent
8	80	
7	70	lower leaves wilted
6	60	
5	50	lower half of plant wilted
4	40	
3	30	top leaves still turgescent
2	20	
1	< 5	plant completely wilted
0	0	as 1; leaves show necrosis

To obtain information on possible interactions, correlations were calculated for the parameters mentioned. To get an indication of the value of the scores for wilting and recovery, as criteria of drought tolerance, the scores from the first set of experiments were compared with the available scores for drought tolerance shown in the variety list (Anonymous, 1965-1985, Stegemann & Schnick, 1982). The score for drought tolerance of potato varieties in the Dutch variety list is based on data from field trials. Visual observations on the foliage of a potato crop on water deficient soils form the main basis for this score. In addition yield data on potato varieties grown under dry conditions are used (Van der Woude, personal communication).

The reproducibility of the test was also studied. Selection, based on recovery scores of the varieties was simulated. The score for recovery of the selected varieties was compared with the known scores for drought tolerance of the varieties. On the basis of these comparisons the value of the greenhouse test for potato breeding is discussed.

Results

Genotypic differences in wilting and recovery

In all experiments large differences between genotypes in scores for wilting and recovery were found. Table 2 summarises the results of the experiments with varieties. The mean scores for wilting based on the data from the experiments with varieties, varied from less than 1.5 (for the variety Provita) to more than 4.0 (for Katahdin). The mean scores for recovery ranged from less than 3.4 (for Première, Ultimus and Resy) to more than 8.0 for Katahdin. A similar spread of scores was observed for the 46 progenitors.

An analysis of variance, with the mean scores for recovery from experiments IV, V and VI, showed significant differences between genotypes and between experiments ($p < 0.05$). The least significant different between genotypes in scores for recovery is 2.3 at the 5 per cent level. In all experiments Katahdin had the highest scores for wilting and recovery.

The group of 46 progenitors included three clones, derived from the same cross. It is interesting to note that the scores for recovery of these clones were 2.1, 3.7 and 6.7 respectively, indicating that within a progeny large differences may occur.

Reproducibility of the results

A measure of the reproducibility of the scores for wilting or recovery is the correlation coefficient between data of separate experiments. For wilting the correlation coefficients (r) were 0.46, 0.50 and 0.44 for the 52 varieties and 0.61, 0.41 and 0.51 for the 46 progenitors. For recovery the correlation coefficients (r) were 0.78, 0.72 and 0.74 for the 52 varieties and 0.69, 0.62 and 0.79 for the 46 progenitors. Rank correlation coefficients were similar to these correlation coefficients.

Table 2. Mean scores for wilting, just before rewatering, and recovery, after rewatering, of 47 varieties in experiments I, II and III.

Name of variety	Score in variety list for		Score for	
	maturity	drought tolerance	wilting	recovery
Arran Banner	6.5	9	3.0	7.4
Katahdin	7.0	9	4.7	8.2
Kennebec	6.0	9	4.0	7.5
Pimpernel	3.5	9	2.9	6.0
Olalla	6.0	9	4.0	6.8
Up to date	5.0	9	2.9	6.2
Bintje	6.5	8	2.0	4.4
Ackersegen	3.5	8	2.6	6.9
Ari	7.0	8	2.9	6.0
Allerfr. Gelbe	7.0	8	2.4	4.4
Désirée	5.5	8	2.9	6.0
Eba	4.5	8	3.2	6.3
Palma	7.0	8	2.9	6.3
Elkana	4.5	8	3.4	7.6
Mara	3.5	8	2.1	5.7
Prevalent	4.0	8	2.9	6.4
Procura	4.0	8	3.2	6.1
Voran	4.5	8	2.3	4.9
Pentland Dell	7.0	8	2.6	6.0
Cardinal	5.0	8	2.5	6.1
Astarte	3.0	7	1.7	6.4
Ehud	7.5	7	1.7	3.7
Mentor	5.0	7	2.1	5.7
Multa	3.5	7	1.6	5.6
Pansta	5.5	7	3.3	5.9
Resy	7.5	7	1.8	3.2
Irene	5.0	7	2.9	6.2
Ambassadeur	4.0	7	2.3	6.7
Prominent	4.5	7	2.4	6.5
Rector	5.0	7	2.0	5.1
Ultimus	6.5	7	1.7	3.3
Amera	4.0	7	2.7	6.4
Amigo	4.0	6	2.3	5.6
Aurora	7.0	6	1.9	4.6
Element	6.0	6	1.8	4.4
Ostara	8.0	6	2.2	5.2
Maris Peer	7.0	6	3.2	5.6
Ulster Premier	8.0	5	2.2	4.2
Provita	7.5	5	1.4	3.8
Woudster	5.0	5	2.7	4.5
Ysselster	6.5	5	2.3	5.2
Bevelander	6.5	5	2.5	5.6
Veenster	6.0	5	1.7	4.1
Saturna	5.5	5	2.3	4.7
Civa	9.5	5	2.6	3.4
Krostar	7.0	5	1.6	5.2
Première	8.5	5	1.9	3.3

* Anonymous, 1965-1985, Stegemann & Schnick, 1982.

9 = very drought tolerant/very early   1 = very drought sensitive/very late

The reliability of the results

The correlation coefficients between the scores for wilting or recovery of 47 varieties and the scores for drought tolerance in the variety list were 0.57 and 0.64, respectively (see Figure 1); rank correlation coefficients were r = 0.56 and r = 0.67, respectively.

SCORE FOR DROUGHT
TOLERANCE VARIETY LIST

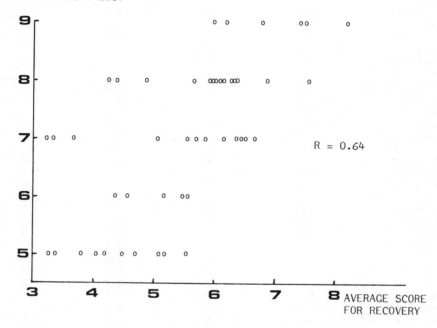

Fig. 1. The relationship between the average scores for recovery of 47 varieties and the scores for drought tolerance in the variety list.

Figure 2 shows the relationship between the scores for recovery and the relative yield on light soils in two dry summers for nine varieties. The relative yield was calculated from yield data of two dry summers, 1976 and 1983*. It appears that the results of the pot test correlated well with relative yields under dry conditions.

---

* Data were kindly supplied by the Government Institute for Research on Varieties of Cultivated Plants (RIVRO).

60

RELATIVE YIELD
UNDER DRY CONDITIONS

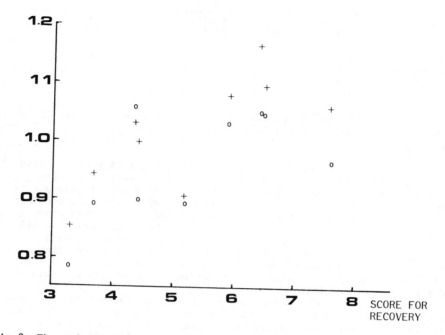

Fig.2. The relationship the score for recovery of nine varieties and the relative yield under dry conditions on sandy soil (o , r = 0.63) or on high peat soil (+ , r = 0.75).

The interrelationship of scores for wilting and recovery and of other plant characteristics

The coefficient of correlation between wilting and recovery was 0.70 for the first and 0.74 for the second series of experiments. The cause of these high r-values is obvious: slightly wilted plants will recover relatively faster and better than severely wilted plants, which have lost turgor for a much longer time. However, some varieties deviate consistently from this trend (see Table 3). These exceptions will be considered in more detail in the discussion.

Table 3. Varieties showing extreme deviations from the general relation between the scores for wilting and recovery.

Variety	Score for wilting	Expected score* for recovery	Observed score for recovery
Multa	1.6	4.3	5.6
Ambassadeur	2.2	5.1	6.7
Woudster	2.7	5.7	4.4
Civa	2.6	5.6	3.4

* Calculation based on the regression of the score for recovery (Y) on the score for wilting (X) : $Y = 2.21 + 1.30 X$ (r = 0.70).

Scores for recovery of the varieties showed a negative correlation (r = −0.52) with the scores for maturity in the variety list; late varieties tend to recover better from the drought stress in the pot test. When the scores for drought tolerance in the variety list are used instead of scores for recovery, the correlation coefficient mentioned above is −0.38 (N = 47). For 80 potato varieties in the Dutch variety list this correlation is −0.34. Apparently the late varieties have a relative advantage in the pot test in that they tend to restore better than earlier varieties. Possibly this is caused by an accelerated ripening of the plants of the earlier varieties under the influence of drought stress under greenhouse conditions.

The correlation coefficients between the scores for wilting or recovery and plant length, plant weight, the difference in plant weight between the "wet" and "dry" treatment, tuber weight and the difference in tuber weight between the "wet" and the "dry" treatment, were negligible for varieties and progenitors as well.

Simulation of selection

In order to assess to what extent recovery scores of the pot test could be used in breeding, different situations, with different selection intensities, were simulated for the group of varieties. These simulated selections were based on the data of one experiment (Table 4) or on the mean scores for recovery of three experiments (Table 5). Tables 4 and 5 give the number of clones, selected on the basis of the score for recovery, having a certain score for drought tolerance in the variety list. Comparing these numbers with the total number of varieties having a certain score, gives an estimation of the chance that (very) drought tolerant genotypes may be missed.

Table 4. Results of simulated selection based on the scores for recovery from one experiment at a time. In experiments I, II and III the same group of varieties was tested.

Score for drought tolerance in the variety list*	Known number of clones per class	Number of clones per class after selection of the best 24 clones based on results of		
		I	II	III
9	6	6	6	6
8	14	9	10	10
7	12	6	6	10
6	5	2	6	7
5	10	1	2	1
total	47	24	24	24

* see subscription Table 2.

Table 5. Results of a simulated selection based on the mean scores for recovery. The mean scores are calculated from the data of experiments I, II and III.

Score for drought tolerance in the variety list*	Number of clones present per class	Number of clones per class after selection of		
		10 best genotypes	10 worst genotypes	24 best genotypes
9	6	4		6
8	14	3	1	11
7	12	3	3	7
6	5		1	
5	10		5	
total	47	10	10	24

* see subscription Table 2

Discussion

A screening method should meet two basic requirements. A prerequisite is that the assessments should give information on the relevant characters. Secondly, the test results should be reproducible. It appears that the (rank) correlation coefficients between data of separate experiments for the scores for recovery are higher than those for wilting, hence the reproducibility of the former scores is higher. Therefore from the viewpoint of reproducibility the recovery seems to be a far better parameter than wilting. Moreover, the correlation coefficient between the score for drought tolerance in the variety list and the score for recovery or wilting is higher for recovery than for wilting.

Significant differences were found between experiments in the potential for recovery after rewatering. Therefore it seems advisable to include in each experiment a number of standard varieties, with known differences in drought tolerance. This allows comparison of results of different experiments. In the results it was stated that in the pot test late varieties tend to restore better than earlier varieties. In view of this it seems advisable to group clones into maturity classes in the pot test.

Simulation studies showed that in a fairly mild positive or negative selection (i.e. 20 per cent of the clones are kept or rejected) not all extreme genotypes will be recognized as such (Table 5). When the best 24 clones are kept, all very drought tolerant varieties will be included and 11 out of 14 clones with an 8 for drought tolerance are in the group of selected clones. This means that 85 per cent of all the varieties with a score for drought tolerance > 7 will be selected. It appears that similar results can be obtained with data of one test only; in other words, replication in time hardly enhances the discriminating ability of the test (Table 4). The major cause of this lack of enhancement is that in the pot test the score for wilting and recovery of certain varieties does not correspond with the score for drought tolerance in the variety list. For example the drought tolerant varieties Bintje and Allerfrüheste Gelbe are classified as "drought sensitive" in the pot test. Therefore the drought tolerance of these varieties depends on (a) mechanism(s) not or only partly evaluated in the pot test. In the Dutch variety list Bintje is classified as drought tolerant, because this variety yields relatively well under dry conditions. The high yielding capacity of Bintje under dry conditions might be due to the potential for resumption of tuber growth, known as second growth, when the water supply is restored.

In the results it was mentioned that a number of varieties deviate from the general relation between scores for wilting and recovery (Table 3). These exceptions could reflect different mechanisms of drought tolerance. It could be speculated that the ability to maintain photosynthesis under dry conditions is expressed in the wilting score. The potential for recovery could be a measure of the speed and the extent of plant growth resumption after a stress period. Very drought tolerant varieties possibly combine both slow wilting, under dry conditions, with fast plant growth resumption after a stress period. On the grounds of the results of these experiments only speculations can be made on the existence of different components for drought tolerance. The following experiments attempt to analyse backgrounds of genotypic differences for wilting and recovery.

It is concluded that the pot test described in this paper is a relatively easy, reliable and cheap technique for screening rather large numbers of clones for drought tolerance. It is expected that the test can be applied successfully when the material is divided into classes according to maturity.

References

Anonymous, 1965–1985. 40e-60e Beschrijvende Rassenlijst voor Landbouw-gewassen (40th-60th Dutch descriptive variety list of agricultural crops). Leiter – Nypels, Maastricht, the Netherlands.

Jones, M.M., Turner, N.C., Osmond, C.B., 1981. Mechanisms of drought resistance. In: "The physiology and biochemistry of drought resistance in plants". ed. L.G. Paleg and D. Aspinall. p. 15–37.

Levitt, J., 1972. Responses of plants to environmental stresses. Academic press, New York, 697 pp.

Loon, C.D. van, 1981. The effect of water stress on potato growth, development and yield. American Potato Journal 58: 51–69.

Loon, C.D. van, 1985. Drought, a major constraint in potato production and possibilities for screening for drought resistance. This contribution.

Miedema, P., 1984. Aanpassing aan ongunstige milieufactoren. SVP-Berichten 20: 20-27.
Stegemann, H., Schnick, D., 1982. Index 1982 Europaïscher Kartoffelsorten. Mitteilungen aus der Biologischen Bundesanstalt für Land- und Forstwirtschaft Berlin - Dahlem Heft 211, 219 pp.
Vos, J., 1985. Research on water relations and stomatal conductance in potatoes. I. An introduction to concepts, techniques and procedures. This contribution.

Acknowledgements

The authors are grateful to Prof.Dr. A.T. van der Wal for providing the data from the experiments I, II and II.

The authors wish to thank Dr. L.M.W. Dellaert and Dr. P. Miedema for critically reading the manuscript, Mr. J. Post for statistically analysing the data and Mr. P.J. Groot, Mr. H. Visser and Ing. G.J. Brommer for their valuable technical assistance.

Virus resistance

MECHANISMS OF VIRUS TRANSMISSION IN RELATION TO BREEDING FOR RESISTANCE
TO POTATO LEAFROLL VIRUS IN POTATO

D. Peters

Department of Virology, Agricultural University, Wageningen

Abstract

Resistance to potato leafroll virus is the result of a complex of
several interacting factors. They are the suitability of the host to
function as a source for the virus, the replication potential of the
virus in the host, and the ability of the host to support colonization
of aphids. A study of these factors may lead to a better understanding
of the resistance to leafroll virus in potatoes.

Introduction

So far, the breeding for resistance to potato leafroll virus (PLRV)
has met limited success. The resistance in potatoes to natural infection
with leafroll virus is studied by assessing the number of infected plants
in a clone. This resistance shows a pattern of continuity between clones
in respect of the number of infected and uninfected plants within each
clone (Davidson, 1973). Infection of a plant is the result of interactions
between the component host, aphid and virus.

Potato leafroll virus, like all members of the luteovirus group to
which also barley yellow dwarf virus and beet mild yellowing virus
belong, is uniquely a phloem tissue virus. It replicates in this tissue
and its occurrence seems to be restricted to it. To transmit the virus
from one plant to another, the aphid has to acquire the virus by inges-
tion of sap from the phloem and to introduce it there again after circulation through the vector. Transmission of this virus can occur only
between those plants which are colonized by the vector. The aphid has to
select the suitable host plant in the field. If a plant is resistent for
some reason to selection by a dispersing aphid, no colonies or only small
ones may develop. If the plant allows the aphid to establish, but is
nutrionally inferior, the aphids grow less rapidly and produce relatively
few offspring resulting again in reduced aphid numbers on the plant. On
the other hand, if a potato plant is accepted by the aphid as a suitable
host, enormous populations of aphids can result. The breeding of potatoes
for resistance to aphids, especially the green peach aphid Myzus persicae
which is the main vector of PLRV, is one factor that has to be studied
separately. How the suitability of the host to support aphid population
is related to the spread of virus and how it influences the efficiency by
which the aphids transmit the virus are not known. A study by Mndolwa et
al.(1984) showed that resistance of M. persicae colonization was apparently not directly related to PLRV infection. These authors concluded
that breeding for resistance to this aphid is not as promising for PLRV
control as developing PLRV resistent cultivars. Aphid resistance per se
may not provide virus control, but size of aphid populations obviously
affects the virus infection pressure which, in turn affects value of
plant resistance (Bagnall, 1977). In this publication, Harrewijn will
later elaborate the role of the host in the development of aphid popu-

lations in more detail. Susceptibility for PLRV varies with the cultivar. Complete immunity does not seem to exist. Thus each potato cultivar or clone supports the replication of virus, but the extend of replication will differ from clone to clone. This property affects the efficiency by which the aphids acquire and transmit the virus. Besides this, the success of transmission also depends on the ease by which the aphid can reach the phloem, the time required to ingest sufficient virus to make the aphids infective, the dose of virus to infect the plant, and the rate by which it spreads, through the phloem. Virology has not been able to answer and describe quantitatively all these properties in detail. Likewise it is not known which of these properties are essential in developing resistance and how they are inherited.

Experimental

To initiate studies on the factors which govern the efficiency of virus transmission, a study was made on the acquisition of the virus, its inoculation, and the availability of virus for acquisition in the infected host. Usually, the transmission is described as a process, the efficiency of which is determined by the length of the acquisition access period and the length of the inoculation period. Generally, it is accepted that the longer the aphid is allowed to feed on infected plants the higher the chance that the aphid becomes infective, and also the longer the aphids, feeds on a plant the greater the probability that the aphid will transmit. This statement has no quantitative value and does not inform us about the quantitative aspects of acquisition and inoculation. To describe the acquisition in a quantitative way, we have developed the concept of AAP_{50}; that is the period in which 50% of the aphids acquire so much virus that they can at least infect one plant. Likewise, studies have been made on the IAP_{50}. This is the period in which 50% of the infective aphids can inoculate a plant.

In these studies, Physalis floridana was used as source and test plant. The AAP_{50} was determined by placing groups of one day old virus-free aphids (M. persicae) for different periods on infected plants. After these periods the aphids were placed on test plants for 5 or 6 days. In this test period the aphid can infect the test plant after circulation of the acquired virus through the aphid's body. The results (Fig. 1) obtained show that the AAP_{50} is approximately 12 hr; a length which could be expected from the results published. The results show a large

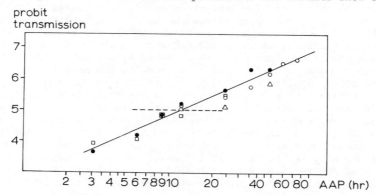

Fig. 1.
Infectivity of aphids in three experiments after different acquisition access period on PLRV infected P. floridana plants. The AAP_{50} is approx. 12 hr.

70

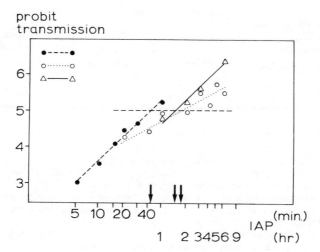

Fig. 2.
The transmission efficiency of infective aphids (M. persicae) in three experiments in short inoculation access period. The IAP_{50}'s found are 45, 96 and 105 min.

deviation in the time required for the individual aphid to acquire sufficient virus to transmit. Some aphids are not able to transmit even after an acquisition period of 48 hr. Assuming that all aphids feed regularily and approximately at the same rate it is concluded that not all aphids are able to acquire virus at the same rate. Since all aphids used are of the same clone and the variation in the virus-acquisition rate by aphids can be explained by a considerable difference in the same age, the concentration of the virus occurring in the separate sieve tubes.

The IAP_{50} was studied as follows. One to 24 hr old aphids were placed on infected plants for 5 days. They were then starved for 2 hr and each aphid was then placed on a test plant for 5, 10, 20, 40, 80, and 160 min. Some aphids were already able to transmit the virus in 5 min (Fig. 2) confirming the results obtained by Klostermeyer (1953) and Holbrook (1978). The IAP_{50} appears to vary from 45 to 105 min in the various experiments. The results can be explained such that the aphids infect a plant at the moment that the stylets penetrate a sieve tube or element. Studies by electronic reading out of stylet penetration, as developed by McLean & Kinsey (1964) and Tjallingii (1985), can perhaps be used to correlate the moment at which the stylet penetrates the phloem and the moment at which the plants becomes infected. The lengths of the period required for an individual aphid to infect the plant varies considerably. This can be explained by the speed at which the aphid can reach the phloem. The period required to reach the phloem spreads from 4 min to at least 3 hr (Tjallingii, 1985). The differences in the IAP_{50}'s found can also be explained by the amount of virus that is injected at the moment of inoculation. When a penetration is not successful with respect of infection or when a dose of virus is too small, more penetrations are required to infect a plant. The size of a dose that can be injected into

Table 1. The LP50 of PLRV in aphids which acquired virus from plants infected by different infection regimes.

Infection regime		Latent period (LP50)	
No of aphids	Infection period	Exp. 1	Exp. 2
1	1	43	41
1	10	53	53
10	10	72	70

a plant will be a function of the amount of virus acquired.

To relate the amount of virus acquired with the efficiency by which the aphids infect a plant, an experiment was designed in which they could acquire virus from plant infected with increasing numbers of aphids and inoculation access periods. One group of plants was infected by one infective aphid in an IAP of 1 hr per plant, another group by 10 aphids in an IAP of 1 hr and a third group by 10 aphids in an IAP of 10 hr. The plants infected in this way showed a gradient of symptoms. Plants with very mild symptoms were found in the first group, those of the second group developed heavy symptoms and those of third group often carried extremely heavy symptoms. Plants with these categories of symptoms were used to study their significance as virus source. To test their efficiency aphids were given an AAP of 24 hr and the length of the LP_{50}, that is the period in which 50% of the aphids make a first transmission, was measured.

The severity of symptoms effects in a negative sense the efficiency of virus transmission. The results (Table 1) showed that the aphids which acquired virus from the first group of plants, transmitted the virus after a short LP_{50}, whereas the aphids which acquired virus from plants infected by 10 aphids in 1 hr or in 10 hr transmitted the virus after LP_{50}'s of approximately 56 and 72 hr, respectively. The lengths of these periods differed statistically longer when plants with severe symptoms were used. Plants with severe symptoms served as poor sources whereas mildly infected plants were rich sources. Although we do not know the amount of virus that can be ingested, the difference found in virus transmission efficiency can be explained by the amount of virus that can be ingested from the plants. It is assumed that the phloem of plants with extremely heavy symptoms is heavily disturbed and, therefore, the sapstream is also severely effected, the aphids can only ingest small amounts of sap and consequently also small quantities of virus.

The sapstream of mildly infected plants will only be slightly effected so that aphids may acquire more virus in a unit to time and will have therefore, a smaller LP_{50}.

Conclusion

Although our experiments and results have a preliminary character, our observations suggest that the factors governing transmission must be analysed further. This may be of interest for the breeding of resistance to PLRV in potatoes. Studies made by breeders show that resistance to PLRV is controlled by a polygenic system. Testing for resistance to PLRV is done either by the release of infected aphids or by mixing the clones with infector plant from which the naturally occurring aphid population spreads the virus. Working in this way the breeder determines a resistance to natural infection of leafroll. The nature of this resistance still remains undetermined and can be the result of interacting factors controlling and regulating the transmission, symptom development, replication of the virus, and the capability to support colonization by aphids. More knowledge about these factors and how they are related is required. It is suggested that the parents to be used in breeding programs must first be analysed with respect to these factors and studied how they are inherited by the offspring.

References

Bagnall, R.H. (1977). Resistance to the aphid-borne viruses in the potato. In: Aphids as virus vectors, pp. 501-526, Eds. K.F. Harris and K. Maramorosch. New York, Academic Press. 559 p.

Davidson, T.M.W. (1973). Assessing resistance to leafroll in potato seedlings. Potato Res. 16: 99-108.

Holbrook. F.R. (1978). Transmission of potato leafroll virus by the green peach aphid. Ann. Entomol. Soc. America 71: 830-831.

Klostermeyer, E.C. (1953). Entomological aspects of the potato leaf roll problem in Central Washington. Techn. Bull. Wash. Agric. Exp. Stations 9, 1-43.

McLean, D.L. & Kinsey, M.G. (1964). A technique for electronically recording aphid feeding and salivation. Nature 202: 462-464.

Mndolwa, D., Bishop, G., Corini, D., & Pavek, J. (1984). Resistance of potato clones to the green peach aphid and potato leafroll virus. Am. Pot. J. 61: 713-722.

Tjallingii, W.F. (1985). Stylet penetration activities by aphids. Thesis, Agricultural University, Wageningen.

THE POSSIBLE USE OF ELISA IN RESISTANCE BREEDING TO VIRUS DISEASES IN POTATO

Xu Pei Wen

guest reseacher at the Foundation for Agricultural Plant Breeding SVP
on leave from the Institute of Vegetable Research, Shandong Academy of
Agricultural Sciences, Jinan, China

Introduction

Virus disease is one of the hazards to potato production in most of the
potato growing countries. Much effort has been put in combating the
disease in various ways among which breeding for resistance is considered
the most effective and economical one. Breeding for resistance is based
on the introduction and identification of different genetical sources of
resistance and the availability of efficient screening techniques.
Knowledge of the nature of resistance serves a theoretical basis for
practical resistance breeding.

In the present paper, problems in resistance breeding to virus
diseases in potato are discussed and the possible development of new
methods for quantitative assessment of the virus content in leaves of
primarily infected potato plants and of the genotypical value of
resistance to virus disease, by employing the enzyme-linked immunosorbent
assay (ELISA) technique, are studied.

Problems in potato resistance breeding to virus diseases

There are a number of viruses which may attact the potato crop. Potato
virus Y (PVY) and potato leafroll virus (PLRV) are the most dangerous
viruses to potato production in most of the potato growing countries,
because of their severe damage to potato growth and yield.

Although it was reported that extreme resistance (or resistance to
multiplication) to potato leafroll virus (PLRV) was found in tetraploid
hybrids of Solanum etuberosum x S. pinnatisectum (15) the resistance to
PLRV commonly used in practice is partial or field resistance, which is
supposed to be of quantitative inheritance, and it is difficult to reach
a high level of the quantitative resistance by accumulating resistance
genes in a conventional breeding program.

For potato virus Y (PVY) the extreme resistance genes originating from
a number of wild Solanum species and S. tuberosum ssp. andigena have been
widely used. Since the inheritance of the extreme resistance is
monogenically dominant, it can be easily incorporated into new varieties.
Moreover, it is strain nonspecific and has so far proved to be relatively
stable (9, 11). However, in spite of the fact that extreme resistance is
available in resistance breeding to PVY, resistance to PVY infection has
also held the attention of the breeders in case the "stable" extreme
resistance is broken.

Besides, attempts to use resistance based on extreme intolerance or
hypersensitivity to both viruses mentioned have also been made by
breeders and researchers (5).

The sources of resistance are limited in cultivated potato and to

74

obtain a high level of resistance and durability the resistance genes available at the present time should be fully exploited, apart from trying to find new resistance genes in wild species.

Generally speaking the more resistance genes a genotype possesses, the higher the resistance level of the genotype will be, and the more difficulties the virus will meet to overcome the resistance. In order to obtain a high level of lasting resistance, combining more types of resistance into one genotype is suggested (14, 16). It is interesting to distinguish different components of the resistance to PVY_N which may differ in their importance for the level of the resistance. We can assume that the level of resistance and the durability could be improved by an optimum combination of a number of resistance genes determining some vital components of resistance. Therefore, the nature of resistance should be studied. If the content of viral antigen in leaves of primarily infected plants can be measured by a suitable technique, like ELISA, studies involving more mathematical analysis on components of resistance, such as resistance to multiplication, translocation and invading of virus, will be feasible.

In potato resistance breeding to virus diseases, the method of assessing genotypical values of resistance should be accurate, cheap and easy to carry out. The traditional field test is still a main method for the assessment of quantitative resistance to potato virus diseases. In this test, plant genotypes are exposed to a certain virus in an infection field. In the following season the plants from tubers harvested from the infection field are assessed for percentage of secondarily infected plants in an evaluation field, or by using other methods. Based on the percentage of secondarily infected plants the genotypical value of resistance is estimated. The variation in percentages of secondarily infected plants caused by the variation in environmental conditions during exposure to the virus is a serious drawback to the method. To obtain a reliable estimate of resistance of varieties or clones, replications in time and space are needed, which makes the testing expensive and time consuming. If the resistance could be estimated by the content of viral antigen, expressed as extinction values in ELISA, in leaves of primarily infected plants the method for assessing the resistance level could be improved considerably. In this procedure plants could be inoculated artificially under controled conditions and tested for the content of viral antigen by means of ELISA a few days after inoculation. In this way the procedure of inoculation and assessment can be finished within a few months in the same season.

ELISA has been used for qualitative detection of potato virus X (PVX), Y (PVY), M (PVM), A (PVA), leafroll (PLRV) and other viruses. However, the possible use of ELISA for quantitative assessment of virus and for estimation of the genotypical value of resistance should be studied. Studies are needed on various aspects, such as the relationship between the virus content and the extintion values in ELISA, between the virus content in leaves and the genotypical value of resistance, the environmental influence on the virus content in primarily infected plants and the interaction between these effects.

Quantitative measurement of virus content in infected leaves.

Clark suggested the use of the ELISA technique in studies on pathogen-host relationships (8). Cardin, L. et al. used the technique in the study on cucumber mosaic virus. In their experiments a good linear relationship (with p=0.01) existed between the extinction values and the log antigen concentration, and a regression line was fitted to the data obtained from dilutions. A "standard" antigen preparation was considered to be used in the estimation of virus concentration of unknown samples (6). In the two experiments carried out in 1985 at the SVP, sap from leaves of 'white Burley' tobacco plants inoculated with PVY^N was diluted in stepwise and then tested by means of ELISA as described by Clark and Adams (7) with the modification given by De Bokx and Maat (4). Healthy tobacco plants were taken as control. Extinction values were measured on a Titertek Multiskan photometer. Fig. 1 shows a close correlation ($r^2=0.94$, $p < 0.01$; $r^2=0.88$, $p < 0.01$) between the relative content of viral antigen (based on the dilution-range)and their extinction values.

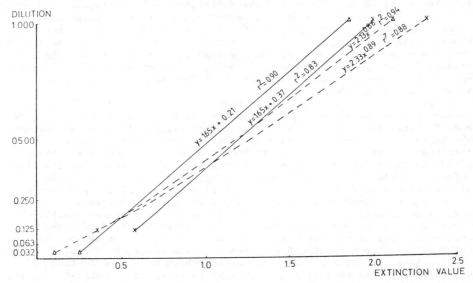

Fig. 1. Regression lines of linear and power functions fitted to the data observed from the dilutions of sap from tobacco leaves infected with PVY^N and their extinction values.
△ test on July 11, x test on July 23, ——— regression line of linear function, ---- regression line of power function

Since the regression line of a power function seems to fit the data better than that of the linear function in the test with PVY^N, the relationship between the relative content of viral antigen from leaves infected with PVY^N and their extinction values might be of a power function. In another experiment the relative content of viral antigen from leaves secondarily infected with potato leafroll virus and their extinction values in ELISA also showed a strong possitive correlation $r^2=0.99$, $p < 0.001$, Fig. 2). A good linear regression line fits the experimental data very well.

From the results it can be deduced that quantitative assessment of the relative content of viral antigen, mainly based on the extinction values in ELISA, is feasible for potato virus Y^N and potato leafroll virus.

Estimation of the level of varietal resistance to PVY^N based on the quantitative measurement of the viral antigen in primarily infected leaves.

Weidemann (18) reported that the extinction values in tubers of plants infected with PVY^N and tested by ELISA differed between varieties but no

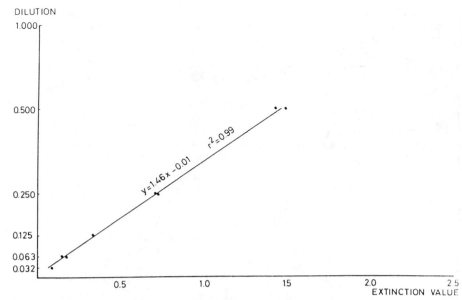

Fig. 2. Regression line fitted to data observed from dilutions of sap from potato leaves infected with potato leaf roll virus (PLRV) and their extinction values in ELISA

correlation was found between the extinction values and PVY susceptibility of potato varieties. Barker (3) found that the concentration of potato leafroll virus (PLRV) antigen, expressed as extinction values in ELISA in leaves of the most resistant genotype, was considerably lower than that in the suceptible one. Van Eijk (17) tested the sprout of tulip plants for mosaic virus by means of ELISA. Differences in the rate of virus multiplication between the susceptible varieties and the less susceptible ones were revealed by comparing the percentages of infected plants. Devergne (10) reported that the level of multiplication of cucumber mosaic virus (CMV) tested by ELISA was much lower in the resistant genotype of muskmellon than in the susceptible one. However, the number of varieties used in the test was limited. There are other studies on the same aspect with pepper and soybean (12, 13).

To study the relationship between the content of viral antigen in potato leaves and the genotypical value of resistance, four experiments were carried out in a greenhouse of the Foundation For Agricultural Plant Breeding (SVP) and a laboratory of the Research Institute For Plant Protection (IPO) in 1984 and 1985 (a). In the experiments plants of 29 potato varieties with different levels of resistance (b) were inoculated with sap with PVY[N] at different plant ages at a temperature of 22-24 °C

(a) detailed description and analysis of the experiments is presented in another paper on the relationship between the relative PVY[N] antigen content in leaves of primarily infected potato plants and the level of varietal resistance to PVY[N].
(b) level of resistance is expressed by the score for resistance given in the Dutch variety list (1, 2), based on field or greenhouse observation.

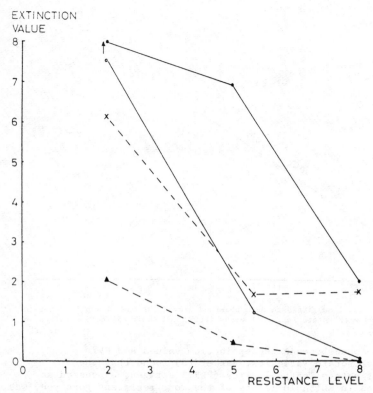

Fig. 3. Mean extinction values of sap from leaves of plants
infected with PVY^N 6 and 8 weeks after planting and assessed
15 days after inoculation
● — ● in inoculated terminal leaflets of plants inoculated
        6 weeks after planting
x — x in inoculated terminal leaflets of plants inoculated
        8 weeks after planting
o — o in uninoculated later al leaflets of plants inoculated
        6 weeks after planting
▲ — ▲ in uninoculated later al leaflets of plants inoculated
        8 weeks after planting

was also considered (Table 1). In the experiments varietal differences in
the extinction values were found to be statistically reliable (p=0.01,
p=0.05), and also a tendency to an increase in extinction values in
leaves as the level of the varietal resistance decreases was revealed in
the first two experiments (Fig. 4). Besides, the results of the
experiments suggest that factors, such as age of plant when inoculated,
time interval between inoculation and testing, and leaf position, may
influence the results of the assessment (Fig. 3, 4). In experiments 3 and
4 a negative correlation between the relative content of viral antigen
expressed by extinction value and the level of varietal resistance is
clearly shown in Fig. 5 (r^2=0.69,p<0.01; r^2=0.86, p<0.01; r^2=0.94,
p<0.01; r^2=0.61, p<0.01).

78

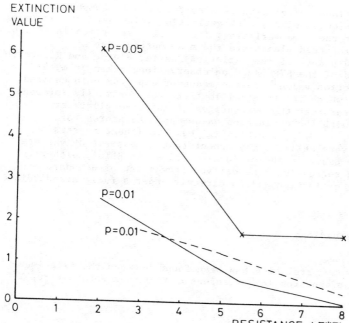

Fig. 4. Mean extinction values in leaves of plants inoculated
8 weeks after planting and assessed 15 days after inoculation
x——x in inoculated terminal leaflets
—— in uninoculated lateral leaflets
--- in the mixture of inoculated terminal leaflets and
uninoculated lateral leaflets

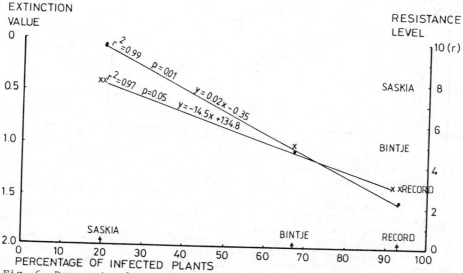

Fig. 6. Regression lines fitted to the data from ELISA, field test
and the Dutch Variety List.
x——x percentage of infected plants and the level of resistance
•——• percentage of infected plants and the extinction values

79

Consequently, a close correlation is also detected by comparison of the percentage of infected plants with extinction values and the varietal resistance level for the same varieties. The results are given in Fig. 6. The percentages of infected plants are the mean percentages of secondarily infected plants of the varieties Saskia, Bintje and Record in the evaluation field of the SVP based on observations taken from 1972 to 1983 (a). The extinction values are the means of sap from uninoculated lateral leaflets of which the terminal leaflets were primarily infected with PVY^N. It is clear that the percentages of infected plants are closely correlated with the extinction values and the scores for resistance ($r^2=0.99$, $r^2=0.97$, p<0.01, Fig. 6). Two linear regression lines fit the data very well. We may expect that assessment of varietal resistance could be based on the extinction values in ELISA, either by comparison with the regression line derived from a set of standard varieties or by the extinction values directly under a fully standardized condition.

Conclusions and discussions

Since a signifcant correlation has been found between the relative content of viral antigen from leaves infected by either potato virus Y^N

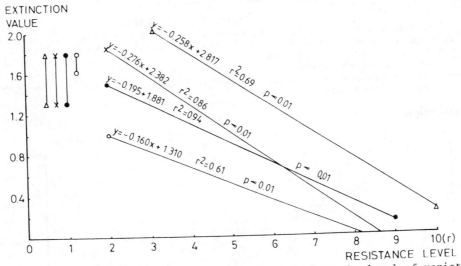

Fig. 5. Regression lines fitted to the data from the level of varietal resistance and the extinction value in ELISA
△——△ mean extinction values from inoculated terminal leaflets tested 13 days after inoculation (exp. 4)
●——● mean extinction values from inoculated terminal leaflets tested 13 days after inoculation (exp. 3)
○——○ mean extinction values from uninoculated lateral leaflets tested 13 days after inoculation (exp. 3)
x——x mean extinction values from uninoculated lateral leaflets tested 16 days after inoculation (exp. 3)
Vertical bars | indicate the least significant difference based on F-LSD test ($\alpha = 0.05$)

(PVY^N) or leafroll virus (PLRV)) and their extinction values in ELISA, the relative content of viral antigen in leaves of an unknown sample can be estimated on the basis of a standard regression line. A negative correlation between the relative content of viral antigen in potato leaves, expressed as extinction values in ELISA and the level of varietal resistance suggests that the level of resistance to PVY^N in some cultivated genotypes of potato can be estimated from the extinction values of sap of leaves of primarily infected plants. However, the expression of the correlation can be influenced by a number of factors which need to be studied further.

Some exceptions to the general correlation were also found (Fig. 5), which might either be an indication of some unknown components of resistance or just of random variation. To measure the exact concentration of virus in leaves, dilution range test with both leaf samples of plant genotypes and purified virus should be made. Besides, leaves to be assessed should be carefully weighted.

Quantitative assessment of the content of viral antigen in leaves of primarily infected plants provides a prerequisite for analytical studies on virus behaviour in host plants, on components of resistance and on the pathogen-host relationship based on quantitative data. The possible estimation of the level of resistance of genotypes based on the extinction values of sap from primarily infected plants can serve as a starting point for developing more efficient methods for assessing resistance.

Consequently mathematical models for unraveling the resistance and predicting the priority of components of resistance could be developed, and further management of resistance in new breeding strategies could be expected. It is interesting to study whether the correlations mentioned above are generally valid for uncultivated potato genotypes, and for other crops, like vegetables and flowers; and for other viruses, like potato leaf roll virus. Much success in the study on resistance and in resistance breeding to virus disease could be achieved if the answers are positive.

(a) data were collected by Wiersema, H.T

Table 1. Potato varieties with different levels of resistance(a) and maturity(b), and the dates of planting, inoculation and assessment in various treatments.

No of experiment	Variety	Score for resistance	Score for maturity	Dates of planting	Dates of inoculation	Dates of assessment	leaves* tested
1	Favorita	8	8	May 1, 15	June 30	July 15	I.U
	Alcmaria	5.5	8	"	"	"	
	Doré	2	9	"	"	"	
2	Bintje	5	6.5	May 1	June 30	July 16	I+U
	Record	3	6.5	"	"	"	
	Désirée	8	5.5	"	"	"	
	Pimpernel	8	3.5	"	"	"	
3	Monalisa	9	7.5	Dec. 17	Jan. 28	Feb. 12,15	I.U
	Blanka	9	7.5	"	"	"	I.U
	Arkula	9	8.5	"	"	"	U
	Gloria	9	9	"	"	"	U
	Producent	8.5	3.5	"	"	"	I.U
	Favorita	8	8	"	"	"	I.U
	Saskia	8	9	"	"	"	U
	Alcmaria	5.5	8	"	"	"	I.U
	Bintje	5	6.5	"	"	"	I.U
	Climax	4	7.5	"	"	"	U
	Record	3	6.5	"	"	"	I.U
	Doré	2	9	"	"	"	I.U
4	Corine	10	9	Jan. 3	Feb. 15	Feb. 28	I.U
	Amera	8.5	4	"	"	"	"
	Favorita	8	8	"	"	"	"
	Ostara	8	8	"	"	"	"
	Alcmaria	5.5	8	"	"	"	"
	Aminca	5	8.5	"	"	"	"
	Sirtema	5	8.5	"	"	"	"
	Prominent	5	4.5	"	"	"	"
	Eersteling	4	9.5	"	"	"	"
	Doré	2	9	"	"	"	"

*   I.U = both inoculated terminal leaflets and the uninoculated lateral leaflets were tested

     U   = only uninoculated lateral leaflets were tested

     I+U = the mixture of inoculated and uninoculated was tested

(a) 9/1: very resistant/very susceptible (Anonymous, 1985)
(b) 9/1: very early/very late (Anonymous, 1985)

References

1 Anonymous, Y^N, -en Bladrolvirusresistentie onderzoek bij aardappel-rassen. (35), (37), (38), (41), (42), 1980-1984. RIVRO, Wageningen, The Netherlands.

2 Anonymous, 1973 and 1985. 48e (60e) Beschrijvende Rassenlijst voor landbouwgewassen. pp. 253 (250-255). RIVRO, Wageningen, The Netherlands.

3. Barker, H., 1984. Studies on mechanisms of resistance in potato to potato leafroll virus. Abstracts of conference papers of the 9th triennial conference of the European Association for Potato Research, Interlaken, 1-6: p. 290.

4 Bokx, J.A. de and Maat, D.Z., 1979. Detection of virus Y^N in tuber with the enzyme-linked immunosorbent assay (ELISA). Meded. Fac. Landbouw. Rijksuniv. Gent 44: 635-644.

5 Brown, C.R., 1977. Breeding for resistance to potato leafroll and Y viruses. Report of planning conference on development in the control of potato virus diseases, CIP, Lima, Peru: 74-80.

6 Cardin, Loïc et al., 1984. Dosage immunoenzymatique (ELISA) du virus de la mosaïque du concombre I. Aspect méthodologique. Agronomie, 1984, 4 (2) 125-135.

7 Clark, M.F., and Adams, A.N., 1977. Characteristics of the microplate method of enzyme-linked immunosorbent assay for the detection of plant viruses. J. gen. Virol. 34: 475-483.

8 Clark, M.F. Immunosorbent Assays in plant pathology. Ann. Rev. phytopathology, 1981. 19 : 83-106.

9 Delhey, R., 1975. Zur Natur der extremen virusresistenz bei der Kartoffel II : Das Y-virus. Phytopathologische Zeitschrift 82 : 163-168.

10 Devergne, J.C. et al., 1984. Dosage immunoenzymatique (ELISA) du virus de la mosaïque du concombre II. Multiplication camparée du virus dans de melons sensibles et resistants Agronomie, 1984, 4 (2), 137-145.

11 Fernandez - Northcote, E.N., 1983. Prospects for stability of resistance to potato virus Y. In: Research for the potato in the year 2000. Proc. of the Int. Congr. in Celebr. of the 10th anniversary of the Int. Potato Center, Lima, Peru. 22-27 Febr. 1982: p. 82.

12 Moore, D.L. et al., 1982. Evaluation of virus contents in soybean by Enzyme-Linked Immunosorbent Assay. Plant disease Vol. 66: 790-793.

13 Marco, S. and Cohen, S., 1979. Rapid detection and titer evaluation of viruses in pepper by Enzyme-Linked Immunosorbent Assay. Phytopathology Vol. 69: 1259-1262.

14 Nelson, R.R., 1984. Strategy of breeding for disease resistance. In: Vose, P.B. & Blist, S.G. Crop Breeding, p. 33-49 Oxford, New York, Toronto, Paris, Frankfurt.

15 Rizvi, S.A.H., 1983. Extreme resistance to potato leafroll virus (PLRV) in seedlings of Solanum etuberosum x S. pinnatisectum (EP) with 4x chromosomes. In: Research for the potato in the year 2000. Proc. of the Int. Congr. in celebr. of the 10th anniversary of the Int. Potato Center, Lima, Peru. 22-27 Febr. 1982.

16 Russell, G.E., 1978. Plant breeding for pests and diseases resistance

17 Van Eijk, J.P., 1983. Componenten-analyse bij de veredeling op resis-tentie tegen mozaïekvirus in tulp. Bedrijfsontwikkeling, 15 (1) 1984: 81-82.

18 Weidemann, H.L, and Wigger, E.A., 1984. The concentration of potato virus Y (PVY) in tubers and leaves of potato plants. Abstracts of Conference papers of the 9th Triennial Conference of the European Association of Potato Research, Interlaken, Switserland, p. 259.

Acknowledgement

The author thanks Prof. Sun Hui Sheng and Dr. J.A. de Bokx for the advice on the study of the subject and Ir. A.G.B. Beekman, Dr. L. Dellaert, Dr. A.B.R. Beemster and Dr. D.E. van der Zaag for the helpful discussions and valuable suggestions on the manuscript.

The author is grateful to Ir. K. van der Woude for kindly supplying information and to Ingrid Schwarz and G. Berends for the typing work and the English corrections. P. Piron, D. Budding, L. Sijpkes and G.J. van Leeuwen are acknowledged for their help in preparing buffers in ELISA, ordering seed potatoes and working on computer.

The Research Institute for Plant Protection IPO is also acknowledged for the permission of using the equipments for ELISA.

VECTOR RESISTANCE OF POTATO TO APHIDS WITH RESPECT TO POTATO LEAFROLL
VIRUS

P. Harrewijn

Research Institute for Plant Protection, Wageningen, The Netherlands[1]

Summary

   Aphids are the most important vectors of persistent viruses in potato.
Both the ability and effectivity to transmit persistent viruses into the
potato crop depend on the species of aphid. A basic difference exists
between aphid resistance and vector resistance. Aphid resistance can be
based upon non-acceptance or upon antibiosis. In both cases, however,
transmission of persistent viruses is possible as long as the aphids take
up phloem sap for a prolonged period of time.
   Vector resistance is only possible when virus is not transmitted for
whatever reason. Such resistance can be based upon the aphids' behaviour
(plant and probing site selection) or can originate in the plant (non-
preference or antibiosis affecting duration of probing and length of
feeding period).
   Examples of types of resistance are presented and possibilities to induce vector resistance are discussed.
   Keywords: PLRV-transmission, host selection, allomones, kairomones,
   antibiosis, phloem sap.

Introduction

   Reduced chances of a plant to become infected with an air-borne virus
can be based on resistance to the virus or to its vector. In potato, leafroll is transmitted by aphids that feed on phloem.
   The aphid life cycle is a theme with many variations. Aphids often have
a primary host, usually a shrub or tree on which they lay hibernating
eggs. The fundatrix that emerges from these eggs gives birth to viviparous
females that can produce spring migrants. These winged morphs fly to
secondary host plants, which mostly are of a limited number of species.
This means that many plant species are in some way resistant to aphids: if
most plants were not resistant to insects the world would not be green.
This natural aphid resistance is often present in wild types or our cultivated plants and may be partly or completely lost during selection towards
modern cultivars. This, for instance, is the case with the resistance of
potato cultivars to the aphid Myzus persicae (Bintcliffe, 1981).
   A basic difference exists between aphid resistance and vector resistance.
Aphid resistance, whether it is based upon non-acceptance or antibiosis,
means that the insects do not multiply on the plants or even die before
they can cause any damage. Vector resistance, which can be specific for a
particular virus, means that this virus is not transmitted (for whatever
reason). Such resistance can be based upon the aphid's behaviour (plant -
and probing site selection) or can originate in the plant (non-preference
or antibiosis affecting duration of probing and length of feeding period).

1. Present address: P.O. Box 9060, 6700 GW Wageningen, The Netherlands

The presence of aphid resistance depends on the following possibilities (Fig. 1):

1. A (potential) host plant can not be found (effect on settling behaviour).
2. After initial probing, the plant is not accepted (effect on host recognition).
3. The plant has been accepted and the aphid feeds, but nevertheless dies or produces little offspring (antibiosis).

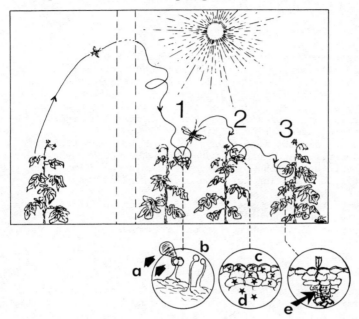

Fig. 1. Defense systems of a plant to aphid attack in three phases: 1: settling, 2: probing, 3: feeding. a = odour and colour perception, b = surface structure (glandular hairs), c = kairomones, d = allomones, e = phloem-bound antibiosis.

In cases (2) and (3) transmission of a persistent virus is possible, as long as the aphids reach the phloem. Only (1) results in both aphid resistance and vector resistance.

Vector resistance also depends on the aphid species or biotype of a species. It should be realized that breeding plants for vector resistance is even more demanding than for aphid resistance.

1. Settling behaviour

Before an aphid reaches the phloem, it has to pass several defensive systems of the plant, the first of which is concerned with host plant recognition. Depending on the length of their flight period and on physical and environmental factors aphids enter a "mood" for landing. Although they cannot fly upwind when air velocity is more than 2-3 m/sec., the air current close to the crop is usually lower and often permits the aphids to alight on a plant of their choice. This choice is the result of silhouette recognition, colour - and sometimes odour perception.

Aphids have eyes with a limited number of facets (usually less than 200) and obtain a low-resolution image of a plant at a distance of a few meters. Nevertheless this seems to be sufficient to terminate flight. Apterous aphids will react also positively to such a silhouette and walk towards a vertical obstacle to climb it.

As soon as the crop is closed, silhouette recognition is more difficult and the chance that aphids will alight becomes lower. This means that host plant resistance can be enhanced by early sewing, resulting in a closed crop before the main flight of a particular aphid species occurs (Harrewijn, 1983). Towards the end of this flight, aphids are attracted by yellow, or at least by light wave lengths above 500 nm. At the same time they reject ultraviolet light. Some potato cultivars flower abundantly with white flowers, reflecting UV-light, as cv. Santé. Compared to cv. Eba we found only 1/3 of winged M. persicae to arrive in a crop of "Santé" at florescence. Of course a potato plant flowers only during a part of the season, but it may be of additional advantage to select for cultivars that have white flowers during the main flights of the most dangerous aphid species.

Many plants have distinctive odours. There is, however, no unequivocal evidence that aphids are attracted by odour when locating their host plants (Chapman et al., 1981). The aphid Nasonovia ribis-nigri for instance, reacts to terpenes in its secondary hosts. The same sensilla are able to detect the alarm pheromone E-β-farnese which can be regarded as a terpene. In this way odours can act as deterrents.

There is at least one documented case that proves that they indeed do. The wild potato Solanum berthaultii has two types of glandular hairs covering its leaves and stems. The so-called A-hairs have a short stalk and a four-lobed head which breaks off when touched and which releases a quick-setting fluid that immobilizes aphids. The B-hairs release E-β-farnesene in amounts sufficient to account for the avoidance of these leaves by wandering apterae of M. persicae (Gibson & Pickett, 1983). It is reported that about 50% of the winged aphids do not alight on the plants, but one can imagine that those that do alight either fly away immediately or start walking will be caught by the sticky A-hairs. Breeding programmes have started to introduce glandular hairs into new cultivars.

2. Host recognition

Most plants do not have such effective glandular hairs as discussed above. After alighting the aphids walk over the upper surface and start probing it with their mouthparts. During this phase there is little or no stylet penetration and the aphid explores the epidermis and exudes saliva on the surface (Pollard, 1977). Once settled, the aphid probes deeply into the plant and eventually can reach the phloem. This process can be stimulated by secundary plant metabolites to which the aphid responds specifically, as it happens with sinigrin. This mustard oil glucoside induces Brevicoryne brassicae, the cabbage aphid, to settle and feed on Cruciferae (Nault & Styer, 1972).

Other metabolites, however, act as allomones, in other words, they deter aphids on their search towards the phloem (Fig. 1). Penetration of the stylets is not necessarily intercellularly, but can be partly intramural or intracellularly. Allomones can be of different chemical nature, from hydrogen cyanides and phenol glucosides to alkaloids. High concentrations of the flavonoids catechin and epicatechin seem to deter Macrosiphum rosae, the rose aphid, from feeding at certain stages of the buds and flowers of rose (Miles, 1985).

Those aphids that feed on plants that contain allomones need detoxication mechanisms. Polyphagous aphids that have a wide range of secundary hosts spend a considerable amount of energy in this respect.

M. persicae is probably the most polyphagous aphid transmitting leafroll into the crop. It prefers host plants that are highly nutritious and it can detoxify a number of allomones. But it cannot cope with all possible toxic substances, especially when they occur in poor hosts. As a result even for this aphid 90% of all vascular plants are not acceptable.

As far as leafroll is concerned, a high level of vector resistance would be obtained when deep probing is interrupted before the phloem is reached. One should realize that only few criteria available to test resistance against aphids can be used for measuring vector resistance (Lehmann & Schliephake, 1983). They are: (a) visual evaluation of superficial probing, (b) number of probes within a certain period of time, (c) uptake of a radioactive tracer such as ^{32}P, (d) measurement of relative growth rate and intrinsic rate of increase, together with (e) reproduction and survival.

Only (a) gives information whether or not the stylets enter the plant, but all other parameters do not discriminate between aphid resistance and vector resistance as one does not know how long sap has been taken up from a sieve element. A better technique to evaluate the degree of vector

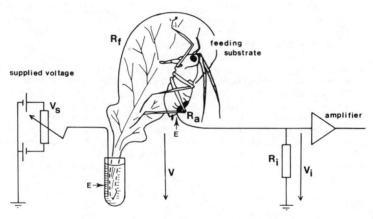

Fig. 2. Recording of an electrical penetration graph (EPG) during stylet penetration. V: circuit potential (= V_s + E), V_s: supplied potential (± 600 mV), E: electrode and other potentials, V_i: signal potential, R_f: plant resistance, R_i: input resistor (according to Tjallingii, 1985a).

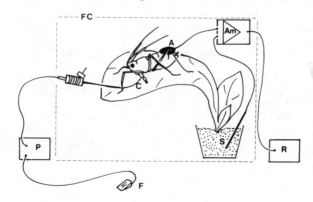

Fig. 3. Stylet cutting with high frequency microcautery. D: radio-source with probe, Am: amplifier, R: recorder for EPG signals, F: foot-switch, FC: cage of Faraday.

resistance in host recognition might be electric recording of penetration behaviour (Tjallingii, 1978) (Fig. 2). If this EPG technique is combined with stylet cutting the question can be answered whether the aphids have been taken up phloem sap (Mentink et al., 1984) (Fig. 3). In this way we could demonstrate that resistance of lettuce selections to Nasonovia ribis-nigri does not prevent the aphids to reach the phloem. Most probably the combination of an allomone in the phloem together with a reduced availability of nutrients form the basis of resistance of lettuce to N. ribis-nigri and M. persicae (Harrewijn & Dieleman, 1984).

According to De Moreno (1983) useful indicators of possible resistance of potato cultivars to M. persicae are the probing behaviour of alatae and the degree of restlessness of the aphid colonies. Alatae probed more frequently on resistant cultivars and although not proved with the techniques mentioned above, it was assumed that less aphids reached the phloem. Consequently the stylet-borne potato virus Y spreads more rapidly on resistant cultivars than on susceptible ones. The nature of this type of resistance is not yet known. The high level of resistance in S. berthaultii is mainly ascribed to the presence of glandular hairs and not so much to the glycoalkaloid composition (Tingey & Sinden, 1982), as plants with high resistance to M. persicae could have a low level of glycoalkaloids.

Massonie (1982) using a somewhat less sophisticated technique as did Tjallingii (1978) suggested that resistance factor(s) operating in peach seedlings express themselves already before the stylets enter the phloem. This does not exclude a possible role of the phloem in this type of resistance.

3. Antibiosis

A. Aphid resistance

The only reliable way to know whether the resistance factor is located in the phloem, is by electrical recording of the probing activities, which can be combined with high resolution microscopy, electron microscopy and stylet cutting. With these techniques we found that the resistance of lettuce selections to the aphids N. ribis-nigri and M. persicae is most probably phloem-bound (Mentink et al., 1984). This, however, does not elucidate the nature of the resistance factor. An indication of the role of nutrition in host plant resistance can be obtained with a system for soilless culture in which aphids can be given access to plants grown on computer-calculated nutrient solutions (Harrewijn & Dieleman, 1984). We could demonstrate that the basis of resistance of lettuce to aphids is most probably the combination of a reduced availability of amino acids and an antibiosis factor (Table 1).

On a partly resistant selection, a high level of nitrogen overruled the natural resistance, resulting in larviposition which was not found under normal circumstances. As far as resistance to aphids is concerned, investigations did not reveal any interaction between temperature, plant genotype and aphid biotype (Eenink & Dieleman, 1982).

B. Vector resistance

Phloem-bound vector resistance to leafroll can be based upon reduced feeding times, with the result that the aphids do not ingest enough virus to become infective. Aphid resistance in no prerequisite for phloem-linked vector resistance. The plant may even be a perfect host for the aphid that nevertheless is inable to acquire virus or to inoculate other plants. Both acquisition time and virus concentration of the phloem sap influence the

Table 1. Development of <u>Nasonovia ribis-nigri</u> on three lettuce selections growing <u>on</u> three different nutrient solutions. M = mortality, $\overline{MRGR}$ = average RGR per column. W_1: 82 µg. x data significantly different from o, or * from x (with $P \leq 0.05$).

me NO_3/l	susceptible			partly resistant			resistant			$\overline{MRGR}$
	$\overline{RGR}$	$W_2(µg)$	M	$\overline{RGR}$	$W_2(µg)$	M	$\overline{RGR}$	$W_2(µg)$	M	
8.25	0.31	1005	o	0.28	817	x	0.10	176	x	0.23
3.00	0.27	686	o	0.24	552	x	0.13	240	*	0.21
1.10	0.22	477	o	0.19	386	x	0.00	dead	*	0.14
$\overline{MRGR}$	0.27 o			0.24 x			0.08 x			

transmissibility of PLRV by aphids (Syller, 1980).

According to Mndolwa et al. (1984) resistance of <u>S. tuberosum</u> gp <u>andiga</u> selections against <u>M. persicae</u> is not related to vector resistance. To give an example: cv. Katahdin is relatively susceptible to <u>M. persicae</u> but highly resistant to PLRV infection. The authors conclude that differences in infection levels among clones were almost entirely due to differences in resistance of plants to infection. It should be noted that resistance of plants to infection is the result of resistance to the virus itself and vector resistance. In the experiments of Mndolwa et al. (1984) the percentage of PLRV infection was compared with the number of aphids on the leaves. Such experiments cannot reveal any difference between virus resistance and vector resistance, although in both cases the phloem may be involved. Concentrations of total foliar glycoalkaloids were positively correlated with colonization by <u>M. persicae</u>, but this merely suggests an indirect relationship of aphid performance and TGA levels to some common determinate which is not measured.

As mentioned earlier about aphid resistance, the only way to know whether vector resistance is phloem-bound, is to use the EPG technique combined with stylet cutting (Mentink et al., 1984). This will provide information on the minimum acquisition time and the proportion of aphids that are actually feeding. Leonard & Holbrook (1978) using an electric monitoring system to record feeding activities of individual aphids found that when aphids were actually feeding, a time of 1.6 min. appeared to be sufficient to acquire enough virus to become infective. Virus inoculation into <u>Physalis pubescens</u> needed slightly longer time: 2.5 min., whereas at least 7.6 min. were needed to transmit PLRV to the potato cv. Russet Burbank.

Prolonged feeding resulted in a higher probability of inoculation after the latent period. The same is true with transmission of PLRV to the indicator plants. A 3-min. pattern of salivation-ingestion resulted in 18.5 % of infected plants: 15 min. of this pattern in 74 % of successful inoculations. A five times longer salivation-ingestion pattern however, does not prove on itself that the aphids were feeding on the phloem much longer.

Uptake may have been from other cells surrounding the phloem. Stylet cutting might have given a definite answer to this question.

These facts indicate that the phloem can have an intrinsic role in vector resistance of potato to aphids. Although obviously a high level of vector resistance is obtained when the aphids are prevented to reach the phloem, a considerable degree of resistance may also be expected when acquisition periods are short due to a deterrent factor in the phloem.

The aphid N. ribis-nigri will start to feed on the phloem of resistant plants, and will withdraw its stylets after a short time. The same happens with Aphis gossypii on resistant muskmelon plants. Grafting does not influence the performance of this aphid on either susceptible or resistant plants. A susceptible part of a graft, be it the scion or the stock, remains just as susceptible. Resistance nor susceptibility are translocated across a graft union and since the aphids find no difficulty in reaching the phloem of resistant plants, it may be that other parts of the phloem such as companion cells produce a phytoalexin acting as a deterrent when they are located by the stylets (Kennedy & Kishata, 1977). Reduced feeding times as discussed above may be the basis of vector resistance to A. gossypii.

The salivation-ingestion patterns produced by Therioaphis maculata while the stylets are inserted in the sieve elements can be interrupted very shortly after the onset of sap uptake (1-2 min.) on a few selections of alfalfa indicating a rapid defense reaction of the phloem cells upon stylet penetration and salivation (Nielson & Don, 1974). These authors mention the possible role of polyphenols in this type of resistance. Certain defensive chemicals like phenols have been recorded to be present in phloem sap and sometimes their presence can be demonstrated because they are sequestered by aphids which in turn use them as defense substances against their natural enemies. This, for instance, is the case with cardenolides imbibed by the aphid Aphis nerii on its host plants Nerium oleander and Asclepia curassavica (Rotschild et al., 1970).

Vector resistance is often specific for an aphid species. This means that when a crop is colonized by more than one aphid species it may be resistant to only one or a few species. In case the phloem is involved, one may expect a difference between resistance to polyphagous and oligophagous aphids. The first category usually prefers highly nutritious plants and does not select for the presence of kairomones, allelochemics used to recognise the host. The second category is less sensitive to nutrition and is often rejected by the presence of a single deterrent, which, however, can turn into a plant substance, which it eventually even prefers (kairomone).

Assessment of vector resistance in potato

Fig. 4 presents a diagnostic step-by-step procedure to assess and locate vector resistance. When vector resistance of a potato selection or cultivar to PLRV should be assessed, it should be done for all aphids known to colonize the crop in the area of investigation. This is because resistance is vector-specific in contrast to resistance to PLRV, which can be independent of the aphids' ability to acquire or inoculate the virus.

Most techniques, published to assess resistance to PLRV may to a certain extent include vector resistance, although the two cannot be further discriminated. Routine exposure to infective aphids in the field is such a method (Wiersema, 1972).

Screening of seedlings in greenhouses by shaking heavily colonized potato shoots above them (Chuquillanqui & Jones, 1980) gives the answer at

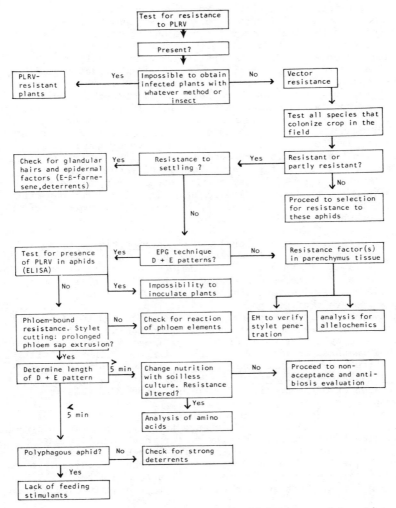

Fig. 4. Diagnostic procedure to establish type and location of vector resistance.

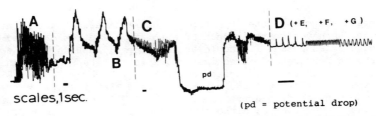

(pd = potential drop)

Fig. 5. EPG patterns showing start of penetration (A,B), stylets in parenchymous tissue (C) with potential drop (puncture of cell wall), presence in phloem (D) with food uptake (D + E). The pattern D + E may proceed for several hours.

an earlier growth stage, but again does not specify the degree of vector resistance. Resistance to PLRV infection is not directly related to aphid resistance (Mndolwa et al., 1984) and the same may hold for vector resistance. How can vector resistance then be established? We can use the following simple rule: resistance to PLRV = vector resistance + resistance to PLRV infection. This means that when the left part of the comparison equals zero, there can be neither vector resistance or resistance to infection. As it is usually more difficult to specifically assess one of the latter two, it is recommendable to use a simple technique for testing resistance to PLRV first. If the answer is positive, the following methods can be used to assess vector resistance:

a. Resistance to settling

Experiments can be done in a glasshouse. Liberate a known number of winged aphids, which are infected with PLRV according to a standardized method, at a certain distance of young susceptible potato plants and do this preferably near the roof of the glasshouse. Count after 1 hr and 6 hrs the number of settled aphids at the upper and lower side of the leaves. The better the plants are accepted, the more aphids will be present at the lower side of the leaves. Count again after 24 hrs and watch whether larvae are deposited. Then remove the aphids (by hand, or with an aphicide) and test for the presence of PLRV as usual. This procedure can be compared with inoculation by caged apterae of the same species. It is possible that a reproducable ratio between the effectivity of winged morphs and caged apterae exists. But one should realize that results may be different in the field, due to differences in host plant recognition and micro-climate.

b. Resistance to host acceptation

Aphids that land on a plant, but do not probe or do so for less than one min. will not reach the phloem and therefore do not transmit PLRV. This type of resistance can simply be assessed by the observation of aphids that either have alighted on the plants or have been put on the leaves. The behaviour of a number of individuals should be followed. Repeated short probes are no problem as long as they are not followed by prolonged insertion of the stylets. When this happens one has to use electric recording of penetration behaviour (Tjallingii, 1978, 1985a, 1985b). The duration of an interrupted D + E pattern gives information on the length of continuous sap uptake (Fig. 5). When this period is more than 5-10 min., the plant could be only partly resistant.

c. Vector resistance to feeding aphids

This is the final phase where vector resistance is possible. When the EPG signals show prolonged feeding and cut stylets produce phloem sap, the only possibility for vector resistance lies in reduced ability to acquire or inoculate PLRV particles. The age of the plants may in itself be a source of resistance, and the availability of PLRV in the lower leaves decreases with age (Syller, 1980). Therefore, aphids must be compared on leaves of the same physiological age. This paper does not discuss whether or not PLRV multiplies in its vectors (Weidemann, 1982) but it should be noted that PLRV transmission is temperature dependent, which is especially important when plants are expected to be only partly resistant to a certain vector. The same holds when one wants to know if vector resistance of aphids that accept the plants is based upon failure to acquire or retain PLRV or to transmit it to other potato plants. Enzyme-linked immuno-

sorbent assay (ELISA) can be adapted for detection and assay of PLRV in
aphids and answers the question whether resistance is expressed in the
acquisition or inoculation phase (Tamada & Harrison, 1981).

When phloem sap is obtained by stylet cutting, it can be analysed both
for the presence of PLRV particles and substances, dissolved in the sap
such as sugars, amino acids, phenolic compounds and many allelochemics.
For analysis of amino acids the amount of sap can be as little as 0.2 -
0.5 μl. Micro-analytical methods can be found among others in Neuhoff,
1973 and Kronberg et al., 1983. More about the technique of cutting
stylets of sucking insects is presented by Fisher & Frame (1984).

References

Bintcliffe, E.J.B., 1981. Resistance to the aphid Myzus persicae (Sulz.)
   in potato cultivars. SROP/WPRS Bulletin 1981 IV. (1): 29-34.
Chapman, R.F., Bernays, E.A. & Simpson, S.J., 1981. Attraction and repul-
   sion of the aphid, Cavariella aegopodii, by plant odours. Journal of
   Chemical Ecology 7: 881-888.
Chuquillanqui, C. & Jones, R.A.C., 1980. A rapid technique for assessing
   the resistance of families of potato seedlings to potato leaf roll
   virus. Potato Research 23: 121-128.
Eenink, A.H. & Dieleman, F.L., 1982. Resistance of Lactuca assessions to
   leaf aphids: components of resistance and exploitation of wild Lactuca
   species as sources of resistance. Proceedings of the 5th international
   Symposium on Insect-Plant Relationships, Wageningen: 349-355.
Fisher, D.D. & Frame, J.M., 1984. A guide to the use of the exuding-stylet
   technique in phloem physiology. Planta 161: 385-393.
Gibson, R.W. & Pickett, J.A., 1983. Wild potato repels aphids by release
   of aphid alarm pheromone. Nature 302: 608-609.
Harrewijn, P., 1983. The effect of cultural measures on behaviour and
   population development of potato aphids and transmission of viruses.
   Mededelingen der Faculteit Landbouwwetenschappen Rijksuniversiteit Gent
   48/3: 791-800.
Harrewijn, P. & Dieleman, F.L., 1984. The importance of mineral nutrition
   of the host plant in resistance breeding to aphids. Proceedings of the
   6th international Congress on Soilless Culture: 235-244.
Holbrook, F.R., 1978. Transmission of potato leafroll virus by the green
   peach aphid. Annals of the Entomological Society of America 71: 830-831.
Kennedy, G.G. & Kishaba, A.N., 1977. Response of alate melon aphids to
   resistant and susceptible muskmelon lines. Journal of Economic Entomo-
   logy 70: 407-410.
Kronberg, H., Zimmer, H.G. & Neuhoff, V., 1983. In: Hirai, H. (editor):
   Electrophoresis. De Gruyter, Berlin/New York: 245-248.
Lehmann, W. & Schliephake, E., 1983. Vergleich von Methoden zur Prüfung
   der Resistenz von Kulturpflanzen gegen Aphiden. Tagesbericht der
   Akademie von Landwirtschaftlichen Wissenschaften der DDR 216: 667-677.
Leonard, S. & Holbrook, F.R., 1978. Minimum acquisition and transmission
   times for potato leafroll virus by the green peach aphid. Annals of the
   Entomological Society of America 71: 830-831.
Massone, G., 1982. Actography of Myzus persicae on susceptible or resis-
   tant peach seedlings. Proceedings of the 5th international Symposium
   on Insect-Plant Relationships, Wageningen: 427-428.

Mentink, P.J., Kimmins, F.M., Harrewijn, P., Dieleman, F.L., Tjallingii, W.F., van Rheenen, B. & Eenink, A.H., 1984. Electrical penetration graphs combined with stylet cutting in the study of host plant resistance to aphids. Entomologia experimentalis et applicata 35: 210-213.

Miles, P.W., 1985. Dynamic aspects of the chemical relation between the rose aphid and rose buds. Entomologia experimentalis et applicata 37: 129-135.

Mndolwa, D., Bishop, G., Corsini, D. & Parek, J., 1984. Resistance of potato clones to the green peach aphid and potato leafroll virus. American Potato Journal 61 (21): 713-722.

De Moreno, C., 1983. Cambios en el compartamiento de Prueba de formas aladas de Myzus persicae (Sulzer) en Siete Variedades de Papa y la disemninacion de los virus PVY y PLRV. Revista Colombiana de Entomologia 9: 1-4; 31-36.

Nault, L.R. & Styer, W.E., 1972. Effects of sinigrin on host selection by aphids. Entomologia experimentalis et applicata 15: 423-437.

Neuhoff, V., 1973. In: Neuhoff, V. (editor): Micromethods in Molecular Biology. Springer: Berlin/Heidelberg/New York: 85-147.

Nielson, M.W. & Don, H., 1974. Probing behaviour of biotypes of the spotted alfalfa aphid on resistant and susceptible alfalfa clones. Entomologia experimentalis et applicata 17: 477-486.

Pollard, D.G., 1977. Aphid penetration of plant tissues. In: Harris, K.F. & Maramorosch, K. (editors): Aphids as Virus Vectors. Acad. Press New York etc.: 105-118.

Rothschild, M.J., von Euw, J. & Reichstein, T., 1970. Cardiac glycosides in the Oleander aphid Aphis nerii. Journal of Insect Physiology 16: 1141-1145.

Syller, J., 1980. Transmission of potato leafroll virus by Myzus persicae (Sulz.) from leaves and plants of different ages. Potato Research 23: 453-456.

Tamada, T. & Harrison, B.D., 1981. Quantitative studies on the uptake and retention of potato leafroll virus by aphids in laboratory and field conditions. Annals of applied Biology 98: 261-276.

Tingy, W.M. & Sinden, S.L., 1982. Glandular pubescence, glycoalkaloid composition, and resistance to the green peach aphid, potato leafhopper, and potato flea beetle in Solanum berthaultii. American Potato Journal 59: 95-106.

Tjallingii, W.F., 1978. Electric recording of penetration behaviour by aphids. Entomologia experimentalis et applicata 24: 521-530.

Tjallingii, W.F., 1985a. Electrical nature of recorded signals during stylet penetration by aphids. Entomologia experimentalis et applicata 38: 177-186.

Tjallingii, W.F., 1985b. Membrane potentials as an indication for plant cell penetration by aphid stylets. Entomologia experimentalis et applicata 38: 187-193.

Weidemann, H.L., 1982. Zur Vermehrung des Kartoffelblattrollvirus in der Blattlaus Myzus persicae (Sulz.). Zeitschrift für angewandte Entomologie 94: 321-330.

Wiersema, H.F., 1972. Breeding for resistance in de Bokx, J.A. (editor): Viruses of potatoes and seed potato production. Centre for Agricultural Publishing and Documentation, Wageningen: 174-187.

DEVELOPMENT OF PARENTAL LINES FOR BREEDING OF POTATOES RESISTANT TO
VIRUSES AND ASSOCIATED RESEARCH IN POLAND

M.A. Dziewońska
Institute for Potato Research, Młochów, 05-832 Rozalin, Poland

Summary
    In the Laboratory of Breeding for Virus Resistance of the Institute for
Potato Research work is done in the following directions: 1. Development
of parental lines with multiple resistance to PVX (gene Rx), PVY (gene
Ry), PVS (gene Ns), PVM (genes GM or Rm in suitable genetic background),
and PLRV (polygenic resistance). 2. Study of inheritance of resistance to
PVY, PVM and PLRV. 3. Search for more effective methods of screening for
virus resistance.

Keywords: breeding, potatoes, resistance, viruses: PVY, PVA, PVX, PVS,
PVM, PLRV, screening methods.

Development of parental lines with multiple resistance to viruses
    The breeding of parental lines resistant to viruses was started in
Poland more than 30 years ago. At first was introduced the resistance to
infection with viruses PLRV and PVY. In the course of the time the
resistance to infection to PVY was replaced by extreme resistance, and
resistance to other viruses important in Polish conditions was added. The
work was done at first in the Breeding Station at Stare Olesno, and since
1966 it is being continued in the Laboratory of Breeding for Virus
Resistance at Młochów in the newly organized Institute for Potato
Research.
At present we attempt to introduce a high level of resistance to six
viruses, which are of major importance in Poland (Table 1).

Table 1. Types of resistance to viruses used in parental line breeding.

Virus	Type of resistance	Responsible genes	Source
PVX	Extreme resistance	Rx	adg acl
PVY+PVA	Extreme resistance	Ry	sto
PVY	High resistance to infection	unknown	sto chc
PVS	Hypersensitivity	Ns	adg
PVM	Necrotic reaction + resistance to infection	Rm+unknown	meg+tbr
	High resistance to infection	Gm+unknown	grl
PLRV	Resistance to infection	unknown	tbr chc

    The breeding is done both at the tetra- and at the diploid levels. At
the tetraploid level the first objective is to obtain clones with multiple
resistance to all the six viruses. In addition the clones are expected to
produce a high tuber yield with starch content 16-18%, and to be mid early
to mid late. This is necessary if the breeders have to utilize them as
parents in their crossing programmes, or try to develop from them new
cultivars directly without farther crossing. At the diploid level the
objective is similar - to combine resistance to viruses with yield and
quality. However in this material we try also to obtain clones homozygous

for resistance genes, and to introduce new sources of resistance. The best diploid clones will be used for crosses with tetraploids to obtain parental lines multiplex for resistance genes.

At both levels of breeding we found no difficulties with the introduction of extreme resistance to PVX and hypersensitivity to PVS. The generally utilized screening methods i.e. mechanical inoculation of first year seedlings and then the confirmation of resistance by grafting, we find satisfactory.

It is different with extreme resistance to PVY. Our most advanced tetraploid clones with the gene Ry are all male sterile. In younger materials we have also male fertile clones, but till now we did not succeed to transfer this character to diploids although we utilized both principal methods: crosses with phu and anther culture. At the diploid level we have identified, until now only a high level of resistance to infection. It was found in materials originating as well from tbr, as from chc. The results of screening for resistance are often difficult to interpret.

In the reaction to PVM we found three types of resistance. At the tetraploid level we found only the resistance to infection. This could be significantly improved by intercrossing clones showing a certain level of resistance. In last years we began to utilize the necrotic reaction to PVM governed by the gene Rm from meg (Ross, Jacobsen 1976, Dziewońska, Ostrowska 1977, Dziewońska et al. 1978) combined with the resistance to infection from tbr. The first clones with such combination of genes grown for three years in field conditions with a high presure of PVM infection were infected in 37%. By mechanical inoculation in the greenhouse the clone with the gene Rm were infected in 0 to 40% and cv. Certa in 100%. After additional backcrosses with resistant tbr the level of resistance was further improved. Now it is possible to select clones which are not infected with PVM either naturally in the field or after mechanical inoculation in greenhouse conditions (Butkiewicz 1985).

At the diploid level we use only high resistance to PVM discovered in grl (Dziewońska, Ostrowska 1978, Waś et al. 1980). The resistant clones do not became infected with PVM either in field cconditions or by mechanical inoculation in the greenhouse. This resistance is already introduced into the tetraploid level by a 4x x 2x cross. We started to evaluate this material.

Our clones with high level of resistance to infection with PLRV are the results of thirthy years work. The original crosses were made between best Polish breeding clones, and clones obtained from GDR, or well known hybrids of Baerecke. The best selections were intercrossed. Later we started to combine the resistance to PLRV with other resistances (Butkiewicz 1984). Knowing the pedigree of the best resistant clones obtained in our laboratory, and basing on published data on the origin of PLRV resistant materials of other breeding units, we conclude, that all descend from the same sources. Therefore at the diploid level we use not only dihaploids descending from PLRV resistant tbr clones, but also PLRV resistant chc clones. We hope to enlarge in this way the genetic background of the resistance.

After many years of work, with the intention to combine various sources of resistance to viruses, we reached the following situation on the tetraploid level: out of 83 clones grown in the field in 1985, which are already fairly well evaluated, 3 seem to posses resistance to all the 6 viruses, and 36 to four or five of them. In 1985 it is being attempted to make two principal types of crossings:
1. resistance to PVX, PVY, PVA, PVS, PLRV x resistance to PVX, PVY, PVA, PVM, PLRV

2. resistance to PVX, PVY, PVA, PVS, PLRV x resistance to PVM, PLRV

The Polish breeders are interested first of all in resistance to PLRV,
PVY and PVM. We hope that clones with multiple resistance to the above
3 viruses carrying the gene Rm will be supplied to them within next
five years.

At the diploid level until 1983 the breeding was conducted on a much
smaller scale. In 1984 for the first time 14 thousand first year seedlings
were grown in the field. In this material we hope to identify clones with
multiple resistance to viruses (Fig. 1.).

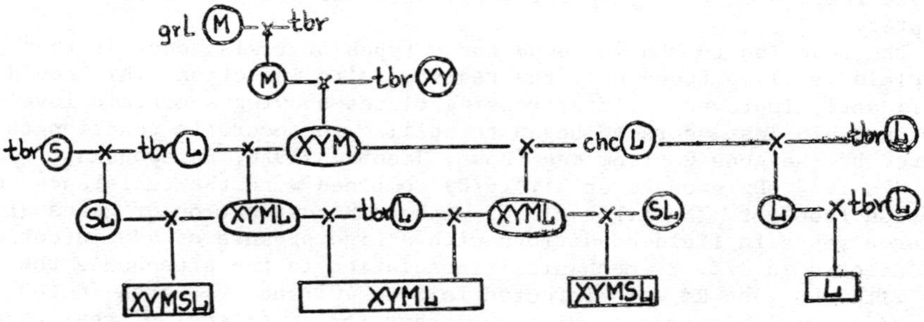

Fig. 1. Course of crossings at diploid level. ◯ clones resistant to given
viruses (X = PVX, Y = PVY, S = PVS, M = PVM, L = PLRV) ▭ clones tested
for resistance to given viruses.

Study of inheritance

As above mentioned at the diploid level we obtained only high
resistance to PVY. It is rather difficult to distinguish it from extreme
resistance due to the gene Ry. When we started breeding at the diploid
level we obtained some tbr dihaploids, which seemed to be extremely
resistant. These clones were used in all programmes in which we wanted to
introduce the resistance to PVY. They were crossed with grl and with chc.
Grl is rather susceptible to PVY, and chc is known to posses some
resistance. In the progenies we found that the number of resistant clones
was much smaller than expected, if the resistance was due to one dominant
gene. As in the near future we do not expect to have diploid clones with
the gene Ry, it is important to learn how the available resistance is
inherited. For this purpose we have selected three resistant clones: one
was a tbr dihaploid, the second its hybrid - with a grl hybrid, and the
third - its hybrid with a chc hybrid. All the three clones have a so high
level of PVY resistance that we were unable to distinguish it from extreme
resistance. Those clones have been crossed in various combinations and the
obtained material is now being investigated.

The problem of identification the PVY extremely resistant clones may be important and difficult to solve. It is likely that high level of resistance is often being confused with extreme resistance. In our work we made several mistakes of this type in tetraploids too. At present we rely on a detailed testing of the evaluated clone, and on the progeny test. In mass screening there are no possibilities for such a detailed study.

Soon after the discovery of a high level of resistance to PVM in grl Swiezyński et al. (1981) found, that this resistance is governed by one dominant gene Gm, and by some modifying factors. We found a useful test for a preliminary determination of the presence of the gene. Clones which are not completely infected in 6 weeks after graft inoculation possess the gene Gm.

The inheritance of resistance to PLRV is little understood. We rely on the old information of Cockerham (1943), Feistritzer (1944), and Baerecke (1956) that the resistance is polygenic and probably dominant. We started recently to get some more information on this problem working with diploids. We obtained six progenies in which one resistant parent has been crossed with six other resistant or susceptible parents. In 1985 we started to evaluate the progenies.

Search for more effective screening methods

To reach advances in our breeding we need effective and simple methods of screening for virus resistance. We are trying to improve the known methods, or to find new ones. For instance Waś et al. (1985) have found that for screening the extreme resistant to PVY clones the mechanical inoculation can replace the inoculation by grafting, when the tested plants are inoculated and grown in stable temperature of 27 $^\circ$C.

Most difficult and time consuming is the evaluation of leafroll resistance. To evaluate it in field trials a proper pressure of natural infection with PLRV is needed. Such conditions are not available at Mlochów. Therefore we were using an artificial inoculation by aphids. For improving this method Syller (1985) is studying the effects of temperature on transmittability of the virus by aphids during the acquisition and inoculation period.

The introdduction of ELISA test for detecting PLRV permitted to evaluate the concentration of the virus in infected plants. The results of Barker (1984), and of specialists in CIP (1985) indicate that in resistant genotypes the virus concentration may be much lower than in susceptible ones. Perhaps this may be utilized to develop a new more effective and simpler method of evaluation of resistance to PLRV.

References

Baerecke, M.L., 1956. Ergebnisse der Resistenzzuchtung gegen das Blattrollvirus der Kartoffel. Z. Pflanzenzuchtung 36: 395–412.

Barker, H., 1984. Studies on mechanism of resistance in potato to potato leafroll virus. EAPR Abstr. of Conf. Pap. 9 Trien. Conf. Interlaken Switzerland, 1–6 Juli, p. 90.

Butkiewicz, H., 1984. Synteza ziemniaków odpornych na wirusy. Zesz. Probl. Post. Nauk Roln. 273: 135–148.

Butkiewicz, H., 1985. Postepy syntezy tetraploidalnych ziemiaków odpornych na wirusy w latach 1980–1984. Zesz. Probl. Post. Nauk Roln. – in press.

Cockerham, G., 1943. Potato breeding for virus resistance. Ann. Appl. Biol. 30: 105–108.

Dziewońska, M.A., B. Czech, K. Ostrowska, M. Waś, 1978. Reaction to potato virus M (PVM) of hybrids with gene Rm derived from S. megistracrolobum. EAPR Abstr. of Conf. Pap., Warsaw, Poland, p. 155–156.

Dziewońska, M.A., Ostrowska K., 1977. Necrotic reaction to potato virus M in Solanum stoloniferum and S. megistracrolobum. Phytopathol. Z. 88: 172-179.

Dziewońska, M.A., K. Ostrowska, 1978. Resistance to potato virus M in certain wild potato species. Potato Res. 21: 129-131.

Feistritzer, W., 1944. Moglichkeiten einer systematischen Resistenzzuchtung gegen die Abbaukrankheiten der Kartoffel. Kuhn - Archiv. 60: 347-357.

International Potato Center, 1985. Annual Report CIP 1984, Lima, Peru.

Ross, H., E. Jacobsen, 1976. Beobachtungen an Nachkommenschaften aus Kreuzungen zwischen dihaploiden und tetraploiden Kartoffelformen, Samenzahl, Ploidiestufen, sowie Spaltungs verhaltnisse des Gens fur extreme Resistenz gegen das X-virus (Rx_{acl}). Z. Pflanzenzuchtung, 76: 265-280.

Syller, J., 1985. The influence of temperature on transmission of potato leafroll virus by Myzus persicae Sulz. Potato Res. - in press.

Swiezyński, K.M., M.A. Dziewońska, K. Ostrowska, 1981.
Inheritance of the resistance to potato virus M found in Solanum gourlayi Haw. Genetica Polonica, 22: 1-8.

Waś, M., M.A. Dziewońska, K. Ostrowska, A. Kowalska, 1980. Reaction of Solanum gourlayi Haw. and its hybrids with S. tuberosum to potato virus M (PVM). Phytopathol. Z. 97: 186-191.

Waś, M., M.A. Dziewońska, H. Butkiewicz, 1985. Postepy w metodach selekcji ziemniaków odpornych na wirusy, oraz w diagnostyce wirusów i wiroida wrzecionowatości bulw ziemniaka. Zesz. Probl. Post. Nauk Roln. - in press.

USE OF PARENTAL LINES IN BREEDING POTATOES RESISTANT TO VIRUSES AT ZAMARTE
BREEDING STATION

A. Pawlak

Institute for Potato Research Experimental Breeding Station, Zamarte,
Poland

The breeding of new varieties at Zamarte breeding station has been done
since 1946. So far, 54 table and starch varieties have been introduced to
the production, among them in Poland the first varieties resistant to
nematodes or immune to virus Y. In addition, the station is doing virus
free seed production of its own varieties and breeding lines as well as of
those from another stations of the Institute for Potato Research. Since
1973 the main breeding goal at the station is the production of starch
potato with resistance to virus diseases and <u>Heterodera rostochiensis</u> Ro_1.
About 120,000 seedlings are raised each year. The program of table potato
breeding is being continued on the scale of 30,000 seedlings a year.

The mating at the station follows the rule of complementation of the
traits in parents and is directed towards broadening the genetic
variability thus composing the basis for selection of the desired
phenotypes. In providing of the high and still growing requirements, in
regard to resistances and yield potential, made on new varieties, the
genetic variability existing among varieties and breeding lines is too
narrow. In order to broaden it, since 1968 the parental lines of the
Department of Genetics (GD), immune (extremely resistant) to virus Y, with
high starch content and adapted to water deficient soils, have been
included in the mating program. In addition, the parental lines are little
related genetically and in matings with our own material the heterosis
effect could be expected.

The station receives from the GD several primarily selected clones each
year for the direct selection of varieties.

From 1968 till 1979 the station received 54 parental lines, marked PG,
which were included in the mating program. Their characteristics are
presented above. In recent years the breeding lines resistant to viruses
(PW) or with high starch content (PS), extremely resistant to virus Y, of
high field resistance to PLRV and to late blight in leaves and tubers,
were used. There were 17 such lines.

The number of seedlings obtained from at least one GD parent is
increasing each year. In table 1 the figures are presented for the number
of breeding lines obtained from GD parental lines in comparison to the
total number of lines selected among starch potatoes during 1973-1979. In
relation to the number of seedlings, both groups of breeding lines,
obtained from GD parental lines and the remaining ones, were represented
in preliminary trials (7-9 years of selection) with the same frequency,
0.25 and 0.24 per mil, respectively. Especially useful were parental lines

PG 168, PG 186, PG 189, PG 292, PG 315, PG 341 and PG 423, which occurred
more often in the pedigree of breeding lines tested in preliminary trials.

Table 1. Number of seedlings and advanced breeding lines from GD parental lines, in comparison with all starch potato breeding lines.

Year	First year seedlings			Breeding lines in preliminary trials			
	total (thousands)	from GD parents		from GD parents		remaining	
		number	%	number	%%	number	%%
1973	99	4	4	4	0.90	31	0.30
1974	111	27	22	4	0.10	26	0.30
1975	140	16	13	1	0.06	26	0.20
1976	122	32	26	13	0.40	21	0.20
1977	129	51	39	13	0.20	13	0.16
1978	162	52	48	4	0.07	11	0.10
1979	152	79	52	25	0.30	29	0.40
Total	915	261	28	64	0.25	157	0.24

Year	Advanced breeding lines	
	from GD parents	remaining
1973	–	–
1974	–	2
1975	–	3
1976	3	2
1977	3	–
1978	.	.
1979	.	.

The cause of quicker elimination of the remaining lines and failure to reach the phase of state trials (10 -12 year of selection) was the specialization at Zamarte station in starch potato breeding and connected with this the rejection of all-purpose lines under selection at the time when specialization was introduced. Exceptions were 2 table lines selected from the year 1974 and 3 lines from 1975 (table 1).

In favour of the parental lines used in the breeding was predominance of advanced breeding lines derived from them in the years 1976 and 1977 over the remaining lines. Moreover, all starch breeding lines accepted for state trails in 1985 have the GD parental lines in their ancestry. These are:

        Z 73072 = Z 56580 x PG 168
        Z 73152 = Uran   x PG 168
        Z 73157 = PG 292   x Z 65480
        Z 74947 = PG 315   x Z 52487
        Z 74967 = PG 315   x Z 52487
        Z 75240 = PG 288   x I.62. 109-3

The parental lines used originated from the starch variety Hochprozentige and S. stoloniferum.

Thus the GD material used has come up to expectations from breeding in the direction of high yielding potential and high starch content, combined with resistance to virus diseases and a satisfactory level of the remaining traits. Therefore the material used turned out to be quit complementary to our own breeding material.

During 1980–1984 the mean share of GD parental lines used as parents of seedlings amounted to 44% (table 2).

Table 2. Number of seedlings grown in 1980–1984.

Year	Number of seedlings (thousands)		
	total	of GD parents	- %
1980	159	90	57
1981	156	76	49
1982	99	23	23
1983	97	58	60
1984	93	22	24
Total	604	269	44

Out of material obtained from GD for the direct selection of varieties, two starch varieties were released in 1983: Brda and Bzura.
Brda (Z–PG 285) is medium late, extremely resistant to virus Y, highly resistant to PLRV.
Bzura (Z–PG 295) is late, extermely resistant to virus Y, highly resistant to late blight, and resistant to wart pathotypes.
The material obtained in recent years become infected with virus M and was discarded before the stage where preliminary trials could be made. The breeding material derived from S. stoloniferum transfers fairly high susceptibility to virus M and it should be mated with partners higly resistant to that virus.

In the specialistic direction of starch potato breeding, accepted in 1973, the GD parental lines comprise a stable and valuable component. The remarkable progress achieved in the resistance to virus Y and having the sources of the resistance to production viruses facilitate the production of breeding lilnes resistant especially to PLRV and late blight. The GD parental lines being genetically distinct give the chance of selecting varieties which higher yielding potential and high starch content.

Analytic breeding methods

THE ANALYTIC BREEDING METHOD: POSSIBILITIES FOR POTATO BREEDING

A.E.F. NEELE & K.M. LOUWES

FOUNDATION FOR AGRICULTURAL PLANT BREEDING, SVP, WAGENINGEN,
THE NETHERLANDS

Summary

Since Chase published his scheme for the analytic breeding method
in 1963, much work has been done to make the method operational.
The research and the material produced have now reached the stage where
the final evaluation of the method can start. For this the question of
the relationship between the performance of the 2x-parent and the
4x-progeny is the most important one. The use of the analytical breeding
method in potato breeding depends on this relationship.
In order to predict the performance of the 4x-progeny for complex
characters, like tuber yield, from data of the 2x-parent it is proposed
that these characters should be brought down to basic physiological
processes. Crop-growth simulation models can possibly be used in
prediction. Plans for research are discussed.

KEYWORDS: Solanum tuberosum, potato, diploids, analytical breeding
method, crop-growth simulation models.

Introduction

As in most countries, practical potato breeding in the Netherlands is
done at the tetraploid level. Problems in breeding due to this ploidy
level are the enormous segregation of the parental genotype during
the meiosis, resulting in a scattering of the parental characters in
the progenies and the unreliable predictability of the level and
variation of the characters in the progenies.
The so-called Analytical Breeding Method, published by Chase (1963),
offers a possible breeding scheme to avoid these problems.
A brief description of the scheme is the following:
- reduction of the ploidy level from tetraploid to diploid
- breeding on the diploid level
- returning to the tetraploid level.

A variant adapted to the modern technological possibilities is
described by Wenzel et al. (1979). The modifications of Wenzel and
coworkers mainly concern a further reduction of the ploidy level to
monoploidy and the use of somatic hybridization.

The reduction of the ploidy level

Methods to reduce the ploidy from 4x to 2x are pollination with the
diploid Solanum species S. phureja (Hermsen & Verdenius, 1973) and
anther culture (Mix, 1982; Wenzel & Foroughi-Wehr, 1984). To produce
diploid breeding material from tetraploids both methods are applicable.
Although the methods are fairly simple, production of a large diploid
population is a laborious and costly procedure as is illustrated by the
following example of production of diploids by the Phureja-pollination
method. During spring 1983 diploid populations were induced by
pollinating 22 cultivars with Solanum phureja IVP 35 and IVP 48. Besides
normal fertilization of egg cells by 2n-gametes, these S. phureja

pollinators induce autonomous growth of unfertilized eggs. The hybrid
seeds can be distinguished from the unfertilized ones by the so-called
seed spot, a purple band at the base of the cotyledons, which can be seen
through the seed coat. This character is inherited by the S. phureja
pollinators and the hybrids exhibit at the base of all organs homologous
to leaves this coloration, the nodle band (Hermsen & Verdenius, 1973). In
the autumn of 1983 the botanical seeds were screened for seed spot with a
binocular and in all more than 10,000 spot-less diploid seeds were
obtained. The seeds marked as unspotted of 14 cultivars were sown in seed
trays after breaking dormancy with an aqueous gibberelic acid solution.
Six weeks after emergence the plants were transplanted to pots and
classified in three groups: a) normal plants, b) plants with growth
abnormalities or chlorophyl deficiencies, and c) plants with nodle band.
The results are given in table 1.

Table 1. The number of berries and spot-less seeds per variety after
pollination with S. phureja IVP 35 and IVP 48 and the percentage of
seedlings of diploid populations of 14 varieties cq. progenitors at the
time of transplanting in the catagories died plants, normal plants,
abnormal plants and plants with nodal band.

Parent	Berries	Spot-less seeds	Seeds sown	Percentage of plants at the time of transplanting			
				died	normal	abnormal	nodal band
Amaryl	942	921	300	33	36	19	12
Sinaeda	537	478	300	38	38	21	3
Premiere	159	281	281	35	46	17	2
AM 70-2166	293	440	440	41	35	23	2
AM 70-2115	550	887	450	17	68	11	3
Ve 7086	292	369	369	21	53	19	7
Ve 68265	379	424	424	25	53	15	7
Ve 71105	394	595	595	27	57	11	5
Ve 709	139	284	284	32	55	11	2
Ve 66295	94	225	225	10	70	17	2
Y66-13-610	291	554	450	27	48	24	1
Y66-13-636	425	1249	450	21	22	53	4
W72-22-496	37	114	82	41	38	15	6
W72-38-720	451	352	352	24	56	15	5
Total	4983	7173	5002	27	49	20	4

The same effort was given during the pollination to all 14
varieties/progenitors. An enormous variation in berry set and number of
spot-less seeds per berry was observed between the varieties, ranging
for .78 spot-less seed per berry in W72-38-720 to 3.08 in W72-22-496.
  Of the 5002 seeds sown, only 2424 developed into normal plants.
Besides, due to absence of flowering the population of plants to be
used in a 2x-breeding programme may be reduced further.
  The efficiency of a method not only depends on its results but also on
its costs. The costs to produce the diploids with the 'Phureja method'
consist of growing costs and labour costs. Of each of the 22 cultivars,
12 plants were grown in the greenhouse by the planting-on-brick method
demanding a total planting space of 300 m^2. The total labour costs were

for about three months work, including maintainance of the plants,
emasculating the flowers, pollination, collecting pollen, harvesting
berries and screening seeds for seed-spot and administration. In all more
than 10,000 spot-less seeds were obtained.
For the reduction of the diploid level to the monoploid level the same
methods are used as for diploids. Instead of tetraploid genitors diploid
genitors are used.
   Compared to the induction of diploids even higher costs and less
success are obtained with the reduction to monoploidy (Breukelen et
al., 1975, 1977; Foroughi-Wehr et al., 1977; Jacobsen & Sopory, 1978;
Sopory et al., 1978; Breukelen, 1981).
At present the rate of success, however, is very low and depends of the
parental genotype. Therefore before breeding schemes based on monoploids
will be applicable more research has to be done in this field.

Breeding at the diploid level
   Breeding of potatoes at the diploid level differs in one important
aspect from the traditional tetraploid potato breeding, since this is
aiming for the production of varieties, combining as many good characters
as possible. Breeding at the diploid level in contrast is in fact
breeding of progenitors, and this calls for a different breeding scheme
than this for breeding varieties.
   Instead of combining all possible resistances and quality characters,
the breeding of genitors requires a more diversified approach. The
breeding programme should consist of several subunits, each focusing on a
limited number of characters. The other characters are of less
importance, but should not be neglected.

Apart from the advantage of the more simple genetics at the diploid
level, breeding at this ploidy level offers the possibility of using
the enormous diploid gene pool of Solanum. Not only for transferring
resistance genes, but also to enrich the S. tuberosum genetic background.
In order to attain maximum of heterozygosity in the final breeding
stage it is advisable to keep the different genetic sources apart from
each other.

In a breeding program it is important to know how reliable the
selection of individual clones is, because the selection intensity and
the selection method depend on the selection reliability and the
breeding aim.
   On the tetraploid level research on the possibilities of the early
generation selection has been done by Maris (1962), Davies & Johnston
(1974), Kameke (1975), Tai (1975), Anderson & Howard (1981)
and Brown et al. (1984). All found that selection for yield and tuber
appearance based on a few plants per clone is unreliable. At the diploid
level no work has been done on selection efficiency. Therefore research
work is needed in this field. To make the breeding scheme, data should be
available on characters which can be properly evaluated in early
generations, on the basis of single plants or row selection. Furthermore
must be known on which characters selection should be based on
(replicated) plots and the breeder has to wait until more plant material
per genotype is available.

Returning to the tetraploid level
   How to return to the tetraploid level? At this moment two different
methods can be used, somatic polyploidization and sexual
polyploidization.

Somatic polyploidization by means of colchicine treatment is not to
be recommended for variety breeding. The results of Jahr et al. (1963)
and Rowe (1967) reveal that somatic doubled tetraploids are in general
worse than parent diploids. Although De, Maine (1984) did not find
differences between the diploid and corresponding tetraploids, all were
less good than the tetraploid standard variety used.
    The other method, the sexual polyploidization, is more promising for
potato breeding. Sexual polyploidization involves unreduced gametes
(2n-gametes).
Recently a third method was developed: the somatic hybridization
(Melchers et al., 1978; Barsby et al., 1984; Gressel et al., 1984; Puite
& Roest, in press).

    Since Chase published his scheme much work has been done, especially
in making diploid breeding material with a good agronomical performance.
However the difficulty of selecting diploid clones for agronomical
characters, resistances and consumption quality is that the final goal is
a tetraploid variety. Therefore the question is whether there is a close
relationship between the performance the 2x-parent and the 4x-progeny.
This is the most crucial question for the analytical breeding method. The
relation between the diploid parent and the tetraploid progeny is not
only of importance for the choice of the selection criteria and breeding
goals at the diploid level but also for the prediction of the performance
of the progeny of a certain combination of diploids.
    The relation between the performance of the diploid and its tetraploid
progeny has been studied by Mok & Peloquin (1975), McHale & Lauer
(1981), Veilleux & Lauer (1981) and Schoeder & Peloquin (1983a & b).
For some characters a high correlation between the diploid parents and
the tetraploid progeny was observed.

    For simple inherited characters, like for instance eye-depth and
maturity, the relation between the diploid parent and the tetraploid
progeny is clear. So the genotypic values of diploids can be used for
prediction of these characters in the progeny.
For a more complex character, as tuber yield, the performance of the
progeny could be predicted by the general combining ability (McHale &
Lauer, 1981; Veilleux & Laver, 1981; Schroeder & Peloquin, 1983b). Also
Mok & Peloquin (1975) found that the for populations derived from 4x x 2x
FDR crosses the main source of variance between the offsprings was due to
differences in general combining abilities of the diploid progenitors.
However ranking of the general combining abilities and tuber yield of the
diploid parents did not show a clear correlation. This might be explained
by the low number of diploid 2n-FDR parents used by Mok & Peloquin. But
another important cause might be that the character tuber yield is,
physiologically, no character in itself, but the result of various
processes, and all these processes are not only determined by the
genetical background of the clone but are also influenced by each other
and by the environment (Burton, 1966). For instance the photosynthetic
rate is not only influenced by light and temperature, but also
tuberization of the plant has its influence on the photosyntetic rate,
the rate increases with tuberization (Ku et al., 1977).
    With complex characters the prediction of the 4x progeny by the
genotypic value of the 2x parent might be improved as these characters
are brought down to basic physiological processes. For instance tuber
yield is determined by processes and characters like photosynthesis,
respiration, plant water balance, tuber initiation, leaf area duration,
increase of the total dry matter and harvest index. However, it is

110

pointless to study all the processes and characters seperately, because of the linkage and interaction between the processes and characters. Therefore a description of the processes in their entirety, as in crop growth simulation models, could be of more use.

Plans for future research at SVP

In the first part of this paper two areas of further research have been pointed out, namely the reliability of selection of diploid clones and the relationship between the performance of the 2x-parent and its 4x-progeny. The research at the SVP focuses on these problems.

The reliability of the selection of diploid clones.

In this project six diploid populations, each derived from a tetraploid cultivar or progenitor, will be followed in their agronomical performance for several years. For a number of characters correlation coefficients between years will be calculated for the diploids. The evaluated characters will be common agronomical characters like tuber form, eye depth, skin colour, stolon length, tuber yield, biomass production and tuber appearance.

For a comparison with tetraploid breeding material, the six varieties were intercrossed according to a scheme for an incomplete diallel, resulting in 20 hybrid populations. Comparisons will be made within the ploidy groups between the clonal generations and between the ploidy groups in the same growing season.

In time the project is seen as follows:
-seedlings grown in pots in a screenhouse. The tuber production and number will be determined.
-1st year clones grown in the field, with two plants per clone. The characters to be evaluated are relative growth rate of the leaves, number of stems, harvest index of fresh material, tuber number, tuber characteristics. To avoid aphid infestation, the plants will be harvested in July.
-2nd-4th-year clones grown in the field, with eight plants per clone in four experimental plots of two plants. Two of the plots will be harvested green in July and the other two in September. The evaluated characters are the same as in the first year clones complemented with characters like senescence and under water weight.

The relationship between characters of the diploid parent and its tetraploid progeny.

The second project concerns the relation between the performance of characters at diploid and at tetraploid level.
As diploid parents six different genotypes are used, all forming unreduced gametes according to the FDR-mechanism. Four of the six diploids are also desynaptic. Desynapsis is a reduced pairing of the homologous chromosomes during the first meiotic division and may result in a reduction of recombination. Therefore it is assumed that gametes of these diploid parents will resemble the parental genotype even more than normal FDR-gametes. The six diploid parents will be evaluated thoroughly, and growth and development of these six will be described by crop growth simulation models.

Not only characters that govern growth and production will be evaluated, but also those less influenced by the environment, like eye-depth, tuber form, under water weight and isozyme patterns.

111

The six diploid genotypes were each crossed with six tetraploid varieties or progenitors to make the tetraploid progeny. The crop growth of the tetraploid parents are also described by a simulation model.

It was not possible to make 2x-2x-crosses, resulting in a tetraploid progeny, because the number of diploids forming 2n-FDR eggs is limited and the available genotypes are closely related to each other and to some of the 2n-FDR pollen producers. Comparisons between tetraploid progenies resulting from diploid-diploid combinations with the corresponding diploid parents will be made in the near future, when either new 2n-FDR egg producers are available or the somatic hybridization technique offers possibilities of combining a fair number of diploids.

The resulting 36 populations are represented by 30 genotypes. Each on four replicates of two plants per clone. During the growing season the plots are frequently evaluated for growth of the green material. To get an impression of the tuber production two plots will be harvested in July and two in September.

The crop growth simulation models can indicate on which characters clones differ from each other in the way of tuber production. The deviating process-rates and characters might point out differences in genetical background of the clones. Rearrangement of these characters and rates of processes in the offspring might lead to higher tuber production. So a more detailed information about crop growth and tuber production of the parents might indicate which cross combinatins are most promising.

Final remark

If there is a close relationship between the diploid parents and their tetraploid progeny the analytical breeding method will offer good possibilities for potato breeding and will change it radically. The main objective of breeding lies than in development of the diploid parents and in the prediction of the best combination, rather than in the selection of the best hybrids. Breeders should keep in mind that breeding at the diploid level is in the first place development of progenitors and the breeding programme must be in accordance with this.

During the development of the diploid parents, it is advisable to make seperate stocks, each differing in genetical and/or taxonomical background, like flint and dent in mais. The groups should be intrahybridized during breeding at the diploid level. To avoid inbreeding and to increase the number of favourable genes, intergroup hybridizations can be performed for breeding the parents of the interploidy crosses. This gives the best guarantee for maximum heterozygosity in the ultimate tetraploid hybrid.

If crop growth simulation models can indicate which ploidy crosses are most profitable for tuber yield, it will be worth while to make growth models of the possible parents. Making these crop growth simulation models will be costly and time-consuming, but the costs will probably be less than spending on clones that would never reach the level of a variety. Nevertheless the costs, the use of crop growth simulation models would make this breeding method a real analytical one. The choice of the parents of the interploidy crosses would be based on a thorough analysis of the parents and calculations of the expected mean and variance in their progenies and not, as now, in tetraploid potato breeding where crosses are made only based on experience of the breeder.

References
Anderson, J.A.D. & H.W. Howard, 1981. Effectiviness of selection in

the early stages of potato breeding programmes. Potato Research 24: 289-299.

Barsby, T.L., J.F. Shepard, R.J. Kemble & R. Wong, 1984. Somatic hybridization in the genus Solanum: S. tuberosum and S. brevidens. Plant Cell Reports 3:165-167.

Breukelen, E.W.M. van, 1981. Pseudogamic production of diploids and monoploids in Solanum tuberosum and some related species. Agricultural Research Reports 908: 121 pp.

Breukelen, E.W.M. van, M.S. Ramanna & J.G.Th. Hermsen, 1975. Monoploids (n=x=12) from autotetraploid Solanum tuberosum (2n=4x=48) through two successive cycles of female parthenogenesis. Euphytica 24: 567-574.

Breukelen, E.W.M. van, M.S. Ramanna & J.G.Th. Hermsen, 1977. Parthenogenetic monoploids (2n=x=12) from Solanum tuberosum L. and S. verrucosum Schlechtd. and the production of homozygous potato diploids. Euphytica 26: 263-271.

Brown, J., P.D.S. Calligari, G.R.Mackay & G.E.L. Swan, 1984. The efficiency of seedling selection by visual preference in a potato breeding programme. The Journal of Agricultural Science, Cambridge 103: 339-346.

Burton, W.G., 1966. The potato, a survey of its history and of factors influencing its yield, nutritive value, quality and storage. H. Veenman & Zonen N.V., Wageningen, Holland.

Chase, S.S., 1963. Analytic breeding in Solanum tuberosum L. -a scheme utilizing parthenotes and other diploid stocks. Canadian Journal of Genetics and Cytology 5: 359-364.

Davies, H.T. & G.R. Johnston, 1974. Reliability of potato selection in the first clonal generation. American Potato Journal 51: 8-11.

De, Maine, M.J., 1984. Comparison of rate of photosynthesis and tuber yield of a dihaploid, its chromosome-doubled derivates and parent. Potato Research 27: 251-260.

Foroughi-Wehr, B., H.M. Wilson, G. Mix & H. Gaul, 1977. Monoploid plants from anthers of a dihaploid genotype of Solanum tuberosum L. Euphytica 26: 361-367.

Gressel, J., N. Cohen & H. Binding, 1984. Somatic hybridization of an atrazine resistant biotype of Solanum nigrum with Solanum tuberosum. Theoretical and Applied Genetics 67: 131-134.

Hermsen, J.G.Th. & J. Verdenius, 1973. Selection from Solanum tuberosum group Phureja of genotypes combining high-frequency haploid induction with homozygosity for embryo-spot. Euphytica 22: 244-259.

Jacobsen, E. & S.K. Sopory, 1978. The influence and possible recombination of genotypes on the production of microspore embryoids in anther cultures of Solanum tuberosum and dihaploid hybrids. Theoretical and Applied Genetics 52: 119-123.

Jahr, W., K. Skiebe & M. Stein, 1963. Bedeutung von Valenzkreuzungen fuer die Polyploidiezuechtung. Zeitschrift fuer Pflanzenzuechtung 50: 26-33.

Kameke, K. von, 1975. Untersuchungen zur quatitativen Variabilitaet in Kreuzungsnachkommenschaften der Kartoffel. Heften fuer den Kartoffelbau 19. 58 pp.

Ku, S.B., G.E. Edwards & C.B. Tanner, 1977. Effect of light, carbon dioxide, and temperature on photosynthesis, oxigen inhibition of photosynthesis, and transpiration in Solanum tuberosum. Plant Physiology 59:868-872.

Maris, B., 1962. Analyse van aardappelpopulaties ten dienste van de veredeling. Thesis Agricultural University Wageningen. Pudoc, Wageningen. 208 pp.

McHale, N.A. & F.I. Lauer, 1981. Breeding value of 2n pollen from diploid hybrids and Phureja in 4x-2x crosses in potatoes. American Potato Journal 58: 365-374.

Melchers G., M.D. Sacristan & A.A. Holder, 1978. Somatic hybrid plants from potato and tomato regenerated from fused protoplasts. Carlsberg Research Comminucations 43: 203-218.

Mix, G., 1982. Dihaploide Pflanzen aus Solanum tuberosum Antheren. Landbauforschung Voelkenrode 32: 34-36.

Mok, D.W.S. & S.J. Peloquin, 1975. Breeding value of 2n pollen (diplandroids) in tetraploid x diploid crosses in potatoes. Theoretical and Applied Genetics 46: 307-314.

Puite & Roest, in press. Protoplast technology; somatic hybridization and culture in potato breeding programmes. In "Potato Research of tomorrow". Pudoc, Wageningen, The Netherlands.

Rowe, P.R., 1967. Performance of diploid and vegetatively doubled clones of Phureja-haploid tuberosum hybrids. American Potato Journal 44: 195-203.

Schroeder, S.H. & S.J. Peloquin, 1983. Parental effects for yield and tuber appearance on 4x families from 4x X 2x crosses. American Potato Journal 60: 819.

Schroeder, S.H. & S.J. Peloquin, 1983. Parent-offspring correlations for vine maturity following 4x X 2x crosses. American Potato Journal 60: 819-820.

Sopory, S.K., E. Jacobsen & G. Wenzel, 1978. Production of monohaploid embryoids and plantlets in cultivated anthers of Solanum tuberosum. Plant Science Letters 12: 47-54.

Tai, G.C.C., 1975. Effectiviness of visual selection for early generation seedlings of potato. Crop Science 15: 15-18.

Veilleux, R.E. & F.I. Lauer, 1981. Breeding behaviour of yield components and hollow heart in tetraploid-diploid vs. conventially derived potato hybrids. Euphytica 30: 547-561.

Wenzel, G. & B. Foroughi-Wehr, 1984. Anther culture of Solanum tuberosum. In: I.K. Vasil (Ed.): Cell culture and somatic cell genetics of plants, Volume I. Academic Press Inc., Orlando. p. 293-301.

Wenzel, G., O. Schieder, T. Przewozny, S.K. Sopory & G. Melchers, 1979. Comparison of single cell derived Solanum tuberosum L. plants and a model for their application in breeding programs. Theoretical and Applied Genetics 55: 49-55.

BREEDING OF DIPLOID POTATOES AND ASSOCIATED RESEARCH IN THE INSTITUTE
FOR POTATO RESEARCH IN POLAND

E. Zimnoch-Guzowska

Institute for Potato Research, 05-832 Rozalin, Młochów, Poland

Summary
    In the Laboratory of Genetics of the Institute for Potato Research
diploid potatoes are being developed outstanding in the following
characters: 1. Satisfactory early tuber yield and resistance to various
pathogens, 2. good cooking quality, 3. high starch content, 4.
resistance to late blight, 5. ability to produce unreduced gametes.
Associated research is done in the following areas: 1. inheritance of
resistance to late blight in hybrids originating from ver and mcd, 2.
inheritance of some cooking quality elements, 3. inheritance of starch
content in hybrids originating from chc and ver, 4. influence of
environmental conditions on the production of unreduced gametes.

Keywords: potato, breeding, diploids, inheritance.

Introduction

   Breeding at the diploid level is one of the important activities of
the Department of Genetics and Parental Line Breeding in the Institute
for Potato Research. Work on virus resistance at the diploid level is
done in the Laboratory of Virus Resistance and will be discussed in the
separate paper. Here will be presented the work of the Laboratory of
Genetics. This work has been started over 15 years ago. It has been
inspired by the idea that wild potato species and primitive cultivars
can be a useful source of genetic variation and by the fact that
segregation of characters at the diploid level is simpler what favours
advances in breeding work and genetic analyses. It seems to be of
special importance that at the diploid level it should be much easier to
obtain homozygotes (Hougas, Peloquin, 1958).

Breeding programs

   The work at the diploid level is done in five separate breeding
programs: 1. satisfactory early tuber yield and resistance to various
pathogens, 2. good cooking quality, 3. high starch content, 4.
resistance to late blight, 5. ability to produce unreduced gametes.
The characters mentioned above should be combined with high yielding
ability, good tuber morphology, resistances to pathogens (viruses,
Phytophthora infestans, nematodes etc.). At present the breeding cycle
is realized during 3 to 4 years. In 1985 about 900 genotypes are being
multiplied on plots with 2, 5, 10 or 15 plants. They are harvested
either 14 weeks after planting (early August) or in the autumn.

Table 1. Mean values of some characters of diploids in breeding program
- data from 1984 (range in brackets).

Program	Number of clo- nes	Tuber yield g/hill	Tuber weight g	Starch yield g/hill	Starch content %
early x yield+res.	146	457 (50-960)	31,0 (16-82)	70 (17-134)	14,8 (9,6-21,2)
cooking x quality	60	605 (164-1530)	35,0 (18-64)	82 (28-200)	13,8 (9,7-20,3)
high starch	194	347 (43-1084)	29,1 (8-85)	72 (21-229)	20,7 (14,2-28,7)
res. to late blight	204	458 (55-2025)	32,4 (8-85)	71 (10-294)	16,3 (9,3-25,9)
2n game- tes	70	398 (80-1080)	22,3 (8-37)	53 (13-167)	13,4 (10,1-19,2)

x = harvest 14 weeks after planting

Each year 10 to 15 thousands first year seedlings are being grown in
the field. Many of them pass a preliminary testing in the greenhouse for
resistance to PVX, PVY or P. infestans.

Early tuber yield and resistance to various pathogenes

In 1985 about 300 hybrid clones are being multiplied. They originate
from dihaploids tbr and from chc, grl, phu, yun, ver or mcd in various
combinations. All the clones in this program are extreme resistant to
PVX, field resistant to PVY and resistant to potato wart. Some of them
are also resistant to nematodes, to PVM and to late blight. Preliminary
tests revealed clones resistant to soft rot, to soft and dry rot or to
PLRV.

Good cooking quality

It is being attempted to obtain two types of clones either with
white, floury flesh or with yellow not desintegrating flesh. The program
started two years ago. Among 248 clones from different programs tested
for cooking quality 60 satisfactory selections have been found.

High starch content

In 1985 in this program 235 hybrids are being propagated in the
field. The high starch content in these clones originates from chc, yun
and ver. To get high starch content sib-crosses and back-crosses have
often been done. Most of the material is resistant to PVX and PVY but
only few clones, obtained recently, are resistant to PVM - character of
special importance in the Mlochów area.

Resistance to late blight

In 1985 are grown 2500 first year seedlings and 204 clones
originating from different resistance sources: ver PI 195170, ver CPC
2644, mcd WAC 3220, phu Soliman CCC 1.3, stn x phu hybrids and from dH
of resistant clones and cultivars (Zarzycka, Osiecka, 1983).
In this material the resistance to late blight has been evaluated in
seedling, leaflet and tuber tests under laboratory conditions. The
resistance level is also being evaluated during natural epiphitotics in
the field.
In 1984 the mean level of resistance of 160 multiplied clones was 6.7
(range 9-4) for leaves and 5.1 (range 9-1) for tubers, according to the
scale of resistance 9-1, where 1 means extremelly susceptible. The
standard for leaf resistance cv. Bronka and the standard for tuber
resistance cv. Sokół received the note 6.

The ability to produce unreduced gametes

The aim of this program is to find clones outstanding in the ability
to produce unreduced gametes. From clones of this program the character
will be introduced into other diploids. This program has been realized
for several years. In 1985 about 3300 first year seedlings and 46 clones
are being propagated.
The preliminary criterion of selection is 5% of big pollen grains,
that is over 25 μm in diameter (Mok, Peloquin, 1975). In more advanced
material is being evaluated seed set in 4x x 2x crosses (over 10 seeds
per berry are required) and cytological abnormalities are looked for in
the meiosis.
It seems that most of our clones produce mixed types of 2n gametes
SDR/FDR. Recently one FDR clone was selected.

Associated research

Four research programs are being realized.

Inheritance of some elements of cooking quality (realized by H.
Jakuczun)

Three main elements are being investigated: the colour of the flesh
and the darkening of the fresh and cooked flesh. It is reported that
flesh colour is inherited either monogenetically in presence of
modifying genes (Okuno, 1952; Schick, 1956) or by two major genes
modified by minor genes (Garg et al., 1981). In both cases yellow flesh
is considered to be a dominant. Darkening of the cooked flesh is
reported to be polygenic (Howard, 1974) and dominant (Stuckey et al.,
1969).
Our experimental material consists of progenies derived from
phenotypically extreme or identical forms and from selfed progenies of
various phenotypes.
We intend to estimate the above characters in first year seedlings. In
1985 about 1500 first year seedlings from 15 progenies will be
evaluated.

Inheritance of starch content (realized by E. Zimnoch - Guzowska)

We found in chc and ver high starch content (Sawicka, 1976). At present we have hybrids originating from these sources. It is intended to investigate how the character is inherited and how it is correlated with other characters. In the work at the tetraploid level with tbr material the starch content was found to be polygenic inherited (Schick, 1956; Engel, 1957) with tendency to dominance of the high starch content (Stevenson et al., 1954; Jaszina 1982).

In 1985 about 1050 first year seedlings from 7 progenies are grown in the field. They originated from the following crosses: - two high starch chc hybrids
- two high starch ver hybrids (in both directions)
- high starch chc hybrid with low starch dH tbr
- high starch ver hybrid with low starch dH tbr
- high starch ver hybrid with high starch chc hybrid (two progenies).

The estimations of starch content will be based on underwater weight. It is intended to evaluate also a part of the clones with other methods i.e. Ewers- Grossfeld method.

Inheritance of resistance to late blight (realized by M. Osiecka)

Hybrids originating from ver and mcd are being analyzed. It is attempted to understand the inheritance of resistance in leaves and in the tubers and the relationship between them. We hope also to determine the relationship between resistance and yielding ability.
In 1985 two unselected progenies are grown. In total they consist of 146 clones coming one from resistant ver and the other from resistant mcd hybrids. In addition 880 first year seedlings of 8 progenies are grown. They originate from various matings of selected genotypes of the initial progenies. According to Abdalla (1970) the resistance to late blight in ver is determined by major genes (Vm) with strong effects of a polygenic background. Our preliminary results indicate the presence of major genes in the both initial progenies.

The influence of environmental conditions on the production of unreduced gametes (realized by I. Wasilewicz)

The aim of this investigation is to define environmental factors stimulating the ability to produce male 2n gametes. Veilleux et al. (1981) and McHale (1983) suggested that low temperature is increasing the percentage of big pollen grains and diads.
In 1985 the following factors are being investigated:
- temperature: a) 27 $^{\circ}$C day, 17 $^{\circ}$C night, b) 16 $^{\circ}$C day, 12 $^{\circ}$C night, c) uncontrolled temperature under greenhouse conditions.
- age of plant - the first inflorescence and the inflorescence produced at the end of the flowering period
- potato plants grafted on tomato and plants growing directly from the tubers.
In this experiment 7 diploid clones are being evaluated:
4 own complex hybrids, $H_2$439 (Jacobsen, 1976) and two phu clones 1936-396 and 1936-640 obtained from SVP.
The following criteria are used: a) fertility, b) percentage of big pollen grains (coloured by lactophenol acid fucsine) c) percentage of diads, triads and tetrads (acetocarmin anther squashes method).

References

Abdalla, M.M.F., 1970. Inbreeding, heterosis, fertility, plasmon differen-
    tiation and Phytophthora resistance in Solanum verrucosum Schlechtd.,
    and some interspecific crosses in Solanum. Dyss. Pudoc, Wageningen.
Engel, K.H., 1957. Grundlegende Fragen zu einem Schema für Arbeiten mit
    Inzuchten bei Kartoffeln. Züchter, 27: 98-124.
Garg, K.C., J. Gopal, A.K. Singh, R.K. Birhman, D. Gupta, 1981. Annual
    Scientific Report of Central Potato Research Institute, Indian Council of
    Agricultural Research, Simla p. 3.
Hougas, R.W., S.J. Peloquin, 1958. The potential of potato haploids in
    breeding and genetic research. Am. Potato J. 35: 701-707.
Howard, H.W., 1974. Factor influencing the quality of ware potatoes. 1. The
    genotype. Potato Res. 17: 490-511.
Jacobsen, E., 1976. Cytological studies on diplandroid potato clone and its
    correlation with seed set in 4x x 2x crosses. Z. Pflanzenzüchtung. 77:
    10-15.
Jaszina, J.M., 1982. Geneticzeskije predposylki vyviedienija
    vysokokrochmalnych sortov kartofela. Genetica XVIII, 7: 1135-1143.
McHale, N.A., 1983. Environmental induction of high frequency 2n pollen
    formation in diploid Solanum. Can. J. Genet. Cytol. 25: 609-615.
Mok, D.W.S., S.J. Peloquin, 1975. The inheritance of three mechanisms of
    diplandoid (2n pollen) formation in diploid potatoes. Heredity, 35, 3:
    295-302.
Okuno, S., 1952. Cytological studies on potatoes with some remarks on
    genetical experiments, Part II. Jap. J. Genet. 27: 3-21.
Sawicka, E.J., 1976. Charakterystyka serii Commersoniana Buk. i gatunku
    S. verrucosum Schlechtd z punktu widzenia przydatności tych form dla
    hodowli ziemniaka. Dyss. Instytut Ziemniaka, Bonin.
Schick, R., 1956. Methoden und Probleme der Kartoffelzüchtung. Sber. dtsch.
    Akad. Landwiss. Berl. 5, 29: 1-40.
Stevenson, F.J., R.V. Akeley, J.G. McLean, 1954. Potato utilizaiton in
    relation to variety (heredity) and environment. Am. Potato J. 31:
    327-340.
Stuckey, J.H., R.E. Tucker, J.E. Sheehan, 1964. Cooking qualities of Rhode
    Island potatoes. Am. Potato J. 41: 1-13.
Veilleux, R.E., F.J. Lauer, 1981. Variation for 2n pollen production in
    clones of Solanum phureja Juz. and Buk. Theor. Appl. Genet. 59: 95-100.
Zarzycka, H., M. Osiecka, 1983. Charakteryzowanie róznych źródet odporności
    na Phythophthora infestans. Zagadnienia odporności ziemniaka na choroby
    grzybowe i bakteryjne. Conf. Bonin, p. 109-119.

THE PRESENT STATE OF RESEARCH INTO THE INDUCTION OF APOMIXIS IN POTATO

E. Jongedijk

Agricultural University, Department of Plant Breeding (IvP), Wageningen, the Netherlands

Summary

In relation to the new technology of growing potatoes from true seeds, the possibility of inducing either diplosporic apomixis or aposporic apomixis in potato has recently received considerable attention.
In potato the induction of diplosporic apomixis appears to offer the best prospects. It might be achieved by combining its genetically controlled elements: a strongly reduced chromosome pairing and/or crossing over during megasporogenesis, the formation of FDR 2n-megaspores and embryo sacs and the parthenogenetic development of the unreduced egg cell.
The present state of research into the possibility of inducing diplosporic apomixis in potato at the Department of Plant Breeding (IvP) is broadly outlined.
Keywords: potato, apomixis, apospory, diplospory, 2n-gametes.

Introduction

During the last decade, considerable attention has been paid to the potential use of 2n-gametes (i.e. gametes with the unreduced, somatic chromosome number) in the cultivated potato, *Solanum tuberosum* L., both in relation to the development of alternative and more efficient breeding strategies (Mendiburu et al., 1974; Peloquin, 1982; Hermsen, 1984b) and the new technology of growing potatoes from true seeds (Hermsen, 1977; Peloquin, 1983). As to the latter, in recent years the possibility of inducing either diplosporic apomixis (Hermsen, 1980; Hermsen et al., 1985) or aposporic apomixis (Hermsen, 1980; Iwanaga, 1982) has been comtemplated.
Both apospory and diplospory involve the formation of female 2n-gametes. In apospory 2n-egg formation is a consequence of the absence of megasporogenesis: the unreduced embryo sac develops directly from a somatic (vegetative) cell of the ovule through mitotic divisions. In diplospory 2n-eggs result from 'abnormal' megasporogenesis: the unreduced embryo sac derives from an archesporial (generative) cell of the ovule, but neither reduction in chromosome number (apomeiosis) nor crossing over occurs during megasporogenesis (Rutishauser, 1967). Both in aposporic apomixis and in diplosporic apomixis, fertilization of the secondary embryo-sac-nucleus may or may not be required for the parthenogenetic development of the unreduced egg cells (pseudogamous and autonomous apomixis respectively).

Inducing aposporic or diplosporic apomixis

At the department of Plant Breeding (IvP) an attempt is being made to artificially induce diplosporic rather than aposporic apomixis in potato. This for the following reasons:
1. There are strong suggestions from earlier literature (Powers, 1945; Petrov, 1970; Hermsen, 1980; Asker, 1980; Matzk, 1982) that gametophytic apomixis comprises a number of distinct and genetically controlled elements. In the case of diplosporic apomixis, the elements that can roughly be dis-

tinguished are a strongly reduced chromosome pairing and/or crossing over
during megasporogenesis, the formation of unreduced megaspores and embryo
sac and parthenogenetic development of the unreduced egg cell. As these
elements are genetically controlled and are available in potato their combination within a single genotype appears to be feasible. In the case of
aposporic apomixis the development of a somatic cell of the ovule into an
unreduced embryo sac and parthenogenetic development of the unreduced egg
cell should be combined. In diploid *S. tuberosum-S. phureja* hybrids Iwanaga (1980, 1982) has claimed the occurrence of aposporic initials, that
might be stimulated (chemical treatment, prickle pollination etc.) to
develop into unreduced embryo sacs giving rise to parthenogenetic progeny.
There is, however, strong evidence that this possibility is based upon a
misinterpretation of what should be considered the normal pattern of megasporogenesis in potato (Jongedijk, 1985).
   2. In attempts to induce diplosporic apomixis in potato, the required
study of 2n-gamete formation through meiotic abnormalities may be expected
to provide vital 'spin-off' information to be used in pursuing the 'Analytic Breeding' method in potato.

It should be mentioned that aposporic (and diplosporic) apomixis might
be induced through mutation breeding. The scanty and, in many cases, contradictory knowledge available about the precise genetic control (monoge-
nic-polygenic; dominant-recessive) of apomixis in additional to the generally remote chances of obtaining 'positive' mutations do not (yet) justify
an extensive mutation breeding program.

Diplospory and 2n-gamete formation

In megasporogenesis various abnormal events may lead to 2n-gamete formation. Depending on the genetic consequences, however, only two distinct
types of 2n-gametes can be distinguished: first-division-restitution (FDR)
and second-division-restitution (SDR) gametes. Basically, FDR gametes can
be regarded to originate from an equational division of the entire (i.e.
numerically unreduced) chromosome complement after completion of prophase
I, which may vary in appearance from typically meiotic to almost completely
mitotic (Ramanna, 1983; Jongedijk, 1985). SDR gametes, in a strict sense,
can be regarded to result from chromosome doubling in the haploid nuclei
that result after completion of the first meiotic division (Ramanna, 1983).
In highly heterozygous crops like potato, FDR gametes are expected to preserve a relatively large amount of the favourable heterozygosity and epis-
tasis present in the parental genotype and thus to strongly resemble each
other and the parental clone from which they derive. SDR, in contrast, is
expected to yield a heterogeneous population of 2n-gametes with increased
homozygosity (Mendiburu et al., 1974; Hermsen, 1984a).
   It is obvious that in an attempt to induce diplosporic apomixis in potato the first step is to combine FDR megaspore formation with reduced chromosome pairing and/or crossing over during megasporogenesis. As a matter
of fact it has been shown that the latter actually is a prerequisite for
the formation of FDR megaspores and consequently for any attempts to induce diplosporic apomixis in potato (Jongedijk, 1985).
   Chromosome pairing and/or crossing over during meiosis may be strongly
reduced by the action of mutant synaptic genes (asynapsis and desynapsis)
or as a result of a lack of chromosome homology (i.e. structural differences between genomes). In these cases SDR and reduced gametes, if produced, are expected to be predominantly sterile due to chromosome imbalance,
whereas FDR gametes will be preponderantly balanced and thus functional
(Ramanna, 1983; Jongedijk, 1985). If crossing over does not occur at all,

such FDR gametes are identical and preserve the parental genotype intact like is the case with diplospory. Though the combination of asynapsis (no chromosome pairing and thus no crossing over) with FDR megaspore formation might be preferred, also desynapsis or genome divergency can be sufficient if crossing over is strongly reduced and predominantly restricted to the chromosome ends.

The present state of research at the Department of Plant Breeding (IvP)

In our attempts to induce diplosporic apomixis, it was tried to combine desynapsis with FDR megaspore formation, because genes for asynapsis are still unknown in potato. Several diploid and recently also tetraploid genotypes that combine desynapsis with a moderate level of FDR megaspore formation have been selected.

In diplosporic apomictic plant species chromosome pairing and crossing over is usually lacking to a great extent and apomeiosis (FDR megaspore formation) may range from almost meiotic (semi-heterotypic division and pseudo-homotypic division) to almost mitotic (mitotized meiosis) (Gustafsson, 1935, 1946).

The Ds/ds locus for desynapsis (dsds = desynapsis, Ds. = normal synapsis) that was used is expressed at both micro- and megasporogenesis (Ramanna, 1983; Jongedijk, 1983). Though this locus is characterized by normal pachytene pairing and a falling apart of bivalent chromosomes during diakinesis, preliminary results of genetic analysis indicate a severe reduction in crossing over. Extensive studies of megasporogenesis in the selected genotypes revealed that FDR megaspores are predominantly formed through pseudo-homotypic division.

At present it is investigated whether and how the FDR megaspores formed in the selected genotypes can be induced to develop parthenogenetically and thus give rise to 'apomictic' progeny. Besides this it is tried to maximize the frequency of FDR megaspore formation.

Acknowledgement

I thank Prof. J.G.Th. Hermsen and Dr M.S. Ramanna for their critical comments on the manuscript.

References

Asker, S., 1980. Gametophytic apomixis: elements and genetic regulation. Hereditas 93:277-293.
Gustafsson, A., 1935. Studies on the mechanism of parthenogenesis. Hereditas 21:1-111.
Gustafsson, A., 1946. Apomixis in higher plants. I. The mechanisms of apomixis. Lunds Univ.Arsskr. 42:1-66.
Hermsen, J.G.Th., 1977. Towards the cultivation in developing countries of hybrid populations of potato from botanical seeds. In: Report Planning Conference on the utilization of genetic resources of the potato, 1977. CIP, Lima, Peru, 101-109.
Hermsen, J.G.Th., 1980. Breeding for apomixis in potato: pursuing a utopian schema? Euphytica 29:595-607.
Hermsen, J.G.Th., 1984a. Mechanisms and genetic implications of 2n-gamete formation. Iowa State J. Res. 58:421-434.
Hermsen, J.G.Th., 1984b. The potential of meiotic polyploidization in breeding allogamous crops. Iowa State J. Res. 58:435-448.
Hermsen, J.G.Th., M.S. Ramanna & E. Jongedijk, 1985. Apomictic approach to introduce uniformity and vigour into progenies from true potato seed

(TPS). In: Report XXVI Planning Conference 'Present and future strategies
for potato breeding and improvement', 1983. CIP, Lima, Peru, 99–114.

Iwanaga, M., 1980. Diplogynoid formation in diploid potatoes. Ph.D.Thesis,
University of Wisconsin, Madison, 113 pp.

Iwanaga, M., 1982. Chemical induction of aposporous apomictic seed production. In: Proceedings International Congress 'Research for the potato
in the year 2000'. CIP, Lima, Peru, 104–105.

Jongedijk, E., 1983. Selection for first division restitution 2n-egg formation in diploid potatoes. Potato Res. 26:399 (Abstract).

Jongedijk, E., 1985. The pattern of megasporogenesis and megagametogenesis
in diploid *Solanum* species hybrids; its relevance to the origin of 2n-
eggs and the induction of apomixis. Euphytica 34: In press.

Matzk, F., 1982. Vorstellungen über potentielle Wege zur Apomixis bei Ge-
treide. Arch.Züchtungsforsch. 12:183–195.

Mendiburu, A.O., S.J. Peloquin & D.W.S. Mok, 1974. Potato breeding with
haploids and 2n gametes. In: K. Kasha (Ed.). Haploids in higher plants.
University of Guelph, Guelph, Canada, 249–258.

Peloquin, S.J., 1982. Meiotic mutants in potato breeding. Stadler Genetics
Symposia 14:1–11.

Peloquin, S.J., 1983. New approaches to breeding for the potato of the year
2000. In: Proceedings International Congress 'Research for the potato in
the year 2000'. CIP, Lima, Peru, 32–34.

Petrov, D.F., 1970. Genetically regulated apomixis as a method of fixing
heterosis and its significance in breeding. In: S.S. Khoklov (Ed.).
Apomixis and breeding; translated from Russian by B.R. Sharma. Amerind
Publishing Co. PVT.LTD., New Delhi-New York, 18–28.

Powers, L., 1945. Fertilization without reduction in guayule (*Parthenium
argentatum* GRAY) and hypothesis as to the evaluation of apomixis and
polyploidy. Genetics 30:323–346.

Ramanna, M.S., 1983. First division restitution gametes through fertile
desynaptic mutants of potato. Euphytica 32:337–350.

Rutishauser, A., 1967. Fortpflanzungsmodus und Meiose apomiktischer Blüten-
pflanzen. Protoplasmatologia, Band VI F3. Springer Verlag, Wien-New York,
245 pp.

THE EFFECT OF TEMPERATURE ON DYAD FORMATION IN <u>SOLANUM PHUREJA</u>

M. Wagenvoort

Foundation for Agricultural Plant Breeding SVP, P.O. Box 117, 6700 AC
Wageningen, the Netherlands

Summary
    The effect of temperature on dyad formation in <u>S. phureja</u> was studied
in six genotypes raised in growth chambers at 13°, 17°, 21° and 25 $^{\circ}$C
under a photoperiod of 16 hours and a relative humidity of 80%. Variable
expression of dyad formation was found to occur among anthers from plants
of the same genotype or from different genotypes grown at different
temperatures. A significant effect of genotype, temperature and
interaction between both was found at 5% probability. Based on the average
dyad formation, the genotypes were divided into three groups, namely
'low', 'intermediate' and 'high'. A relatively low temperature favoured
dyad formation in the last group but not in the low and in the
intermediate group. On the contrary, these groups had the highest dyad
frequencies at relatively high temperatures.

Keywords: <u>Solanum phureja</u>, dyad formation, fused spindles, temperature.

Introduction
    In the analytic breeding scheme of potato, retetraploidization of
diploids (2n = 2x = 24) can be accomplished by 4x x 2x or reciprocal
crosses in which the diploids produce 2n pollen or eggs respectively. A
variable expression of the 2n gamete formation will hamper an efficient
use of the 2x clones in practical breeding because with a low percentage
of 2n gamete formation the 4x x 2x matings will result in small seedling
populations. Seed set in 4x x 2x crosses is roughly related to the
frequency of 2n pollen production as reported by Jacobsen (1976 and 1980),
Quinn et al., (1974) and Schroeder & Peloquin (1983). The last two authors
found a non-linear relationship between the number of seeds per berry and
the percentage of 2n pollen. Veilleux & Lauer (1981) found that 2n pollen
formation is a highly unstable plant trait suggesting a strong effect of
environmental factors on 2n gamete formation. It has been claimed by Mok &
Peloquin (1975 [a,b]) that the occurrence of 2n pollen in potato is
controlled by a single recessive gene ps, responsible for 'parallel
spindle' formation at anaphase II of microsporogenesis. Monogenic
inheritance of 2n pollen production in <u>Solanum</u> has been disputed at length
by Ramanna (1974 and 1979) and Veilleux & Lauer (1981). Ramanna suggested
that several environmental factors such as temperature, daylength, age of
the plant, and the effects of the use of insecticides, influenced the
pattern of meiotic abnormalities and consequently the level of 2n gamete
formation in several clones. Veilleux & Lauer were not able to identify
specific environmental factors with a general effect on the expression of
unreduced gamete formation because of the large genotype x environment
interaction component of variance. However, they also stated that the
temperature regimes employed (cool growth chamber: 8.3° - 12.8 $^{\circ}$C; warm
growth chamber: 12.2 - 17.2 $^{\circ}$C and field conditions) were not so extreme
as to allow estimation of the effects of temperature on meiotic processes.
    Veilleux et al., (1982) carried out an anatomical study on four clones
of <u>S. phureja</u> in which they examined cytological disturbances resulting in
2n pollen production. Cross sections of buds showed that during the second

meiotic division fused, tripolar and parallel spindles can occur in adjacent cells of a locule. Regulation of the orientation of the spindles during second meiotic division is apparently controlled on cell level. Variation in frequency of 2n gametes among anthers (61.7 vs 5.6%) and between locules of an anther (73.1 vs 90.0%) was found to occur. Intraclonal variability for frequency of 2n gametes implies an extreme sensitivity of this trait to microenvironmental conditions. It is suggested that the action of a gene or a gene complex controlling spindle orientation at metaphase II of meiosis depends on the extracellular environment. Furthermore, fused, tripolar and parallel spindles could represent decreasing levels of gene action.

McHale (1983) found substantially higher overall frequencies of 2n pollen under cool field conditions (8° - 22 ^{o}C) than in a warm greenhouse (14° - 40 ^{o}C). All the genotypes studied were sensitive to the conditions in the warm environment, in which compared to the cool climatic conditions the percentages of 2n pollen dropped to 50% or below. Therefore, McHale assumed that temperature is a major factor in determining differences in 2n gamete formation. A hypothesis of delayed function of spindle fibers at second division which could result in coorientation on a single metaphase plate has been put forward (McHale, 1983). Low temperature seemed to enhance phenotypic expression by exacerbating the delay.

The opinion of McHale is contradictory to the findings of Stow (1927). The last author observed in a cultivar (2n = 4x = 48) a normal meiosis at low temperatures (15° - 20 ^{o}C) whereas at higher temperatures (25° - 30 ^{o}C) many univalents were formed at metaphase I followed by irregular splitting of the cytoplasm and dyad formation.

To study the effect of temperature on dyad frequency in genotypes of Solanum phureja with different levels of dyad formation, plants were grown in growth chambers at four different temperatures. In this article the effect of temperature on dyad formation in microsporogenesis in six different genotypes of Solanum phureja is described and discussed.

Material and methods
     The six genotypes of S. phureja used in this study were derived from two clones from the adaptation programme of Dr B. Maris, Foundation for Agricultural Plant Breeding, Wageningen. The two clones designated S. phureja '75-1136-1931 and '75-1136-1936 produced 2n pollen and were heterozygous for an interchange as well as for an inversion. The cytological behaviour of these clones will be described elsewhere. In the first inbred generation of the clone S. phureja '75-1136-1936 three genotypes, i.e. 151, 258 and 354 were selected. The other three genotypes, i.e. 632, 633 and 680 were selected in the progeny of the cross S. phureja '75-1136-1936 x S. phureja '75-1136-1931. Three plants per genotype were grown in each of four growth chambers with a relative humidity of 80% and a temperature of 13°, 17°, 21° and 25 ^{o}C during a light period of 16 hours, alternated with a dark period of 8 hours with 9°, 13°, 17° and 21 ^{o}C, respectively. The plants were grown in plastic pots 16 cm in diameter and placed on trolleys in growth chambers, kindly provided by the Department of Horticulture, Agricultural University, Wageningen, the Netherlands. Light intensity was about 50 Watt m^{-2} at the top leaves.

For the meiotic studies flower buds were collected after a photoperiod of six hours and fixed for 48 hours or longer at 5 ^{o}C, in a mixture of 3 parts ethanol and 1 part propionic acid saturated with ferric acetate, and stored at ca -20 ^{o}C for longer periods. Before squashing the anthers were stained with alcoholic - hydrochloric - acid carmine, according to Snow (1963), for 6 hours at 60 ^{o}C. Subsequently they were rinsed in 70% ethanol and squashed in a drop of 45% acetic acid. Three anthers from one or more buds from each plant were examined: in total 9 anthers per temperature class, and a minimum of 200 Pollen Mother Cells (PMCs) per anther. The experimental design was a split plot. For testing equality of two percentages of dyads, triads or tetrads observed at sporad stage of meiosis, a method of Sokal & Rohlf (1969) based on the arcsine transformation was chosen.

Results and discussion
Dyad frequency and nuclear restitution in genotypes grown in an unconditioned greenhouse.

The six genotypes were grown in late summer and autumn in an unconditioned greenhouse in a season preceding that in which the experiment in the growth chambers was carried out. Table 1 presents for each genotype the mean percentages of dyads and for the genotypes 258, 632, 633 and 680 also the frequencies of normal, parallel and fused spindle orientations at second metaphase (M_2) of meiosis. Based on the mean dyad frequency the genotypes could be divided into three groups. The first group included the genotypes 151 and 354 which had a very low or no dyad formation. The second group, containing genotype 258 only, represented an intermediate level of dyad formation. The genotypes 632, 633 and 680 formed a third group composed of clones with extremely high level of dyad formation (range from 77.5% to 96.2 %, calculated from 2488 PMCs). Although nuclear restitution in the genotypes 258, 632 and 680 was found to occur partly by fused spindles at M_2 of meiosis, the frequencies of either fused or parallel or fused + parallel spindles did not correspond well with the percentages of dyads. Furthermore at second prophase no important aberrations were found in the four clones studied. This implies that from these cytological observations alone it could not be concluded whether one or more meiotic nuclear restitution mechanisms were operative in these clones.

Table 1. Frequency (%) of dyads in six genotypes of S. phureja between brackets the ranges, and frequency of normal (N), parallel (P) and fused (F) spindles at second metaphase (M_2) of meiosis in four genotypes, grown in an unconditioned greenhouse.

Genotype	PMCs	Dyads	(range)	M_2			PMCs
				N	P	F	
151	1486	0.3	(0.0- 0.8)	–	–	–	
354	740	0.4	(0.0- 0.7)	–	–	–	
258	844	27.8	(24.2-32.0)	87.5	5.9	6.6	303
632	3041	85.5	(81.9-92.6)	59.9	6.4	33.7	187
633	2488	90.3	(77.5-96.2)	61.3	20.2	18.5	124
680	2244	90.3	(88.1-91.7)	62.4	9.3	28.3	247

Dyad, triad and tetrad frequencies in the six genotypes grown in growth chambers at four different temperatures.

The Figures 1 and 2 present the means and 95% confidence intervals for frequencies of dyads, triads and tetrads over the four temperatures for the six genotypes. The averages for dyad formation in the individual clones are in accordance with the frequencies observed when these clones were grown in an unconditioned greenhouse. The genotypes 151 and 354 again

had extremely low dyad formation over all temperatures. The mean dyad
frequency of genotype 258 was about 14% (Figure 1) representing the
intermediate group. The genotypes 632, 633 and 680 had a mean dyad
frequency of about 65, 67 and 70% respectively (Figure 2). The frequencies
for triad formation over the four temperatures were low (Figures 1 and 2).
The highest triad frequency was found for genotype 680, i.e. 6.1%. A triad
will generate two reduced gametes, each with half the chromosome number of
the diploid, and one unreduced gamete. On the contrary a dyad will result
in two pollen grains with the diploid number of chromosomes. It is clear
that the low frequencies of triads had a small effect on the level of 2n
gamete formation in the six genotypes studied. The 95% means and
confidence intervals for the frequencies of dyads, triads and tetrads over
the six genotypes grown at four different temperatures are presented in
Figure 3. The comparatively low temperatures (13°, 17° and 21 $^\circ$C) favoured
dyad formation whereas the relatively high temperature of 25 $^\circ$C dropped
the mean dyad frequency over the six genotypes to 15%. However,
significant differences for dyad formation were observed among plants of
the same genotype (Tables 2 and 3). For instance for genotype 258 at 13 $^\circ$C
less variation was observed than at 25 $^\circ$C (Table 2). Genotype 633 showed
the largest variation at 17 $^\circ$C (Table 3).

Table 2. Variation in the formation of dyads, triads and tetrads in eleven
plants of genotype 258 of S. phureja at four different temperatures. A
minimum of 600 PMCs per plant was analysed.
Different letters behind the percentages refer to significant differences
at 1% or 5%* probability.

Temperature ($^\circ$C)	Type of sporad	Plant		
		1	2	3
13	dyads	4.3 a	1.8 b	2.1 b*
	triads	1.4 a	0.0 b	1.4 a
	tetrads	94.3 a	98.2 b	96.5 ab
17	dyads	14.7 ab	17.8 a	13.0 b*
	triads	8.5 a	3.5 b	4.2 b
	tetrads	76.8 a	78.7 ab	82.8 b
21	dyads	22.6 a	19.7 a	
	triads	8.2 a	3.7 b	
	tetrads	72.2 a	73.6 a	
25	dyads	17.3 a	21.8 b*	15.7 a
	triads	5.9 a	5.0 a	6.4 a
	tetrads	76.7 a	73.2 a	77.9 a

Table 3. Variation in the formation of dyads, triads and tetrads in nine plants of genotype 633 of S. phureja at three different temperatures. A minimum of 600 PMCs per plant was analysed.
Different letters behind the percentages refer to significant differences at 1% probability.

Temperature (°C)	Type of sporad	Plant		
		1	2	3
13	dyads	88.5 a	84.9 ab	83.2 b
	triads	2.7 a	3.6 a	3.2 a
	tetrads	8.8 a	11.5 ab	13.6 b
17	dyads	64.6 a	87.7 b	79.8 c
	triads	0.5 ab	0.2 a	1.6 b
	tetrads	34.9 a	12.1 b	18.6 c
21	dyads	84.7 a	83.5 a	66.9 b
	triads	4.1 a	4.3 a	9.8 b
	tetrads	11.2 a	12.2 a	23.3 b

In addition the three distinguished groups of genotypes did not respond similarly to temperature. The Tables 4 and 5 show the frequencies of dyads in the genotypes 151, 354, 258 and 632, 633 and 680 respectively, grown at $13°$, $17°$, $21°$ and 25 °C. Small differences in dyad formation were found in the genotypes 151 and 354 at all temperatures (Table 4). Genotype 354 had the highest dyad frequency at 25 °C, viz 3.7%.

Table 4. Frequencies (%) of dyads of the six genotypes of S. phureja grown in growth chambers at 13°, 17°, 21° and $25\ ^\circ$C.

Genotype	Temperature $^\circ$C				Mean
	13	17	21	25	
151	0.4	0.1	0.1	0.0	0.1
354	0.2	0.4	1.0	3.7	1.0
258	2.6	15.1	24.1	18.3	13.7
632	78.1	86.1	72.7	43.7	70.8
633	85.6	78.1	78.8	21.0	66.9
680	61.4	80.5	87.6	25.3	65.1
Mean	29.2	36.4	37.7	14.9	

Table 5. Frequencies (%) of triads of the six genotypes of S. phureja grown in growth chambers at 13°, 17°, 21° and $25\ ^\circ$C.

Genotype	Temperature ($^\circ$C)				Mean
	13	17	21	25	
151	0.4	0.0	0.0	0.0	0.1
354	0.0	0.0	0.4	0.1	0.1
258	0.6	5.2	5.5	5.8	3.8
632	2.6	1.8	2.6	4.5	2.8
633	3.1	0.6	5.8	1.2	2.3
680	9.5	6.5	2.9	6.4	6.1
Mean	1.7	1.4	2.2	2.1	

The 'intermediate' genotype 258 revealed similar trends for dyad frequencies. This genotype also had the highest dyad frequency at the higher temperatures, ranging from 2.6% at 13 $^\circ$C to 24.1% at 21 $^\circ$C (Table 4). The genotypes of the third group designated 'high' showed high dyad formation at relatively low temperatures, viz 13°, 17° and 21 $^\circ$C (Table 5). However, when plants of this group were grown at 25 $^\circ$C, dyad frequency dropped below 45% for genotype 632 and below 25% for the genotypes 633 and 680.

The results presented in the Tables 4 and 5 suggest interaction between genotype and temperature for dyad formation. To test this hypothesis an ANOVA test for dyad and triad formation was carried out, using a split plot design. The results of the analyses of variance are shown in the Tables 6 and 7.

Table 6. Results from analysis of variance for dyad formation in six genotypes of S. phureja, using a split plot design.

Source	D.F.	SS	MS	F	
Replication	2	43.2446	21.6223		
Temperature (T)	3	2634.6677	878.2225	58.727	P < 0.00
Error A	6	89.7252	14.9542		
Genotype (G)	5	40232.3778	8046.4755	772.072	P < 0.00
Interaction (GxT)	15	5226.9594	348.4639	33.435	P < 0.00
Error B	36	427.2985	10.4219		

Table 7. Results from analysis of variance for triad formation in six genotypes of S. phureja using a split plot design

Source	D.F.	SS	MS	F	
Replication	2	1.6729	0.8364		
Temperature (T)	3	31.9727	10.6575	1.291	P < 0.36
Error A	6	49.5024	8.2504		
Genotype (G)	5	1666.9894	333.3978	98.868	P < 0.00
Interaction (GxT)	15	483.4494	32.2299	9.557	P < 0.00
Error B	36	138.2579	3.3721		

Table 6 shows a significant effect of temperature, genotype and the interaction between both upon dyad formation. The effect of temperature on triad formation was not significant (P = 0.36, Table 7).
However both the effect of the genotype and the effect of the interaction between genotype and temperature upon triad formation were significant. From the results of the ANOVA tests it could be concluded that there was no unidirectional effect of temperature on dyad formation in the six genotypes studied.

Nuclear restitution mechanism(s) in the six genotypes grown in growth chambers at four different temperatures.

To elucidate the restitution mechanism, a minimum of 200 PMCs per genotype was analysed cytologically at different stages of meiosis. Special attention was paid to second prophase (P_2) and metaphase (M_2). No abnormalities were found at second prophase except the precocious separation of sister chromatids from some chromosomes in a few cells. These small disturbances however are negligible because they will not give rise to dyad formation. At second metaphase a variable expression in occurrence of fused spindles was observed. The Tables 8 and 9 present the frequencies of parallel and fused spindles at M2 and the frequencies of dyads and triads estimated in anthers from buds collected at the same date. In general the frequencies of fused spindles estimated for the six

Table 8. Results of cytological analyses of M_2 and sporad stage of meiosis in buds collected at the same date from four plants from each of six genotypes of S. phureja. The frequency of parallel and fused spindles is based on a minimum of 200 PMCs and that of dyads and triads on 600 PMCs.

Genotype	Temperature	Frequency (%)			
		Parallel spindles	Fused spindles	Dyads	Triads
151	13	12.7	0.0	1.1	0.4
	17	4.0	0.0	0.1	0.0
	21	9.3	0.0	0.1	0.0
	25	9.6	0.0	0.0	0.1
	mean	10.3	0.0	0.4	0.2
354	13	21.8	0.0	0.1	0.0
	17	12.5	0.0	0.3	0.0
	21	13.5	0.5	0.3	0.1
	25	14.6	1.0	3.6	0.3
	mean	15.7	0.4	1.1	0.1
258	13	14.6	5.8	2.1	1.4
	17	9.5	3.8	17.8	3.5
	21	15.5	12.6	19.7	3.7
	25	18.2	6.0	15.7	6.4
	mean	15.3	6.3	14.5	3.8

genotypes grown at the various temperatures did not correspond with the related frequencies of dyads. Only for genotype 680, when grown at 25 $^{\circ}$C, was there clear correspondence between the occurrence of fused spindles and dyad frequency (Table 9). Cytological analyses of M_2 of genotype 633

Table 9. Results of cytological analyses of M_2 and spard stage of meiosis in buds collected at the same date from four plants from each of six genotypes of S. phureja. The frequency of parallel and fused spindles is based on a minimum of 200 PMCs and that of dyads and triads on 600 PMCS.

Genotype	Temperature	Frequency (%)			
		Parallel spindles	Fused spindles	Dyads	Triads
632	13	13.9	12.6	67.3	3.3
	17	15.8	16.7	85.2	1.1
	21	10.0	35.7	74.6	4.0
	25	8.4	24.6	44.1	4.6
	mean	12.4	19.6	69.6	3.3
633	13	11.8	27.6	84.4	3.2
	17	1.0	71.0	64.6	0.5
	21	6.8	23.2	83.4	4.2
	-	-	-	-	-
	mean	6.0	44.0	77.9	2.7
680	13	7.2	6.4	62.8	10.6
	17	9.8	9.8	77.4	6.7
	21	12.0	13.5	89.6	1.4
	25	7.9	22.6	24.4	7.2
	mean	9.2	12.8	63.6	6.5

grown at 17 °C revealed 71.0% of fused spindles. The mean dyad frequency of 77.9% and the 64.6% dyads at 17 °C (Table 9) in this genotype are more or less in accordance with the extent of fused spindle formation. However, when 633 was grown at 13 ° and 21 °C a high discrepancy was found between dyad frequency and the percentage of fused spindles at M_2 (83.4% vs 23.2% and 84.4% vs 27.6% respectively). Although ranking of the dyad frequencies of the six genotypes was in agreement with that of the frequencies of fused spindles it remained unclear whether fused spindles alone or also another restitution mechanism was operating in these genotypes.

The genotypes 151 and 354 had a mean frequency of parallel spindles of 10.3 and 15.7% respectively with a maximum of 21.8% for genotype 354, when grown at 13 °C. It can be concluded that parallel orientation of spindles at M_2 mostly does not result in the formation of unreduced microspores since in the genotype 151 and 354 the frequencies of parallel spindles were much higher than the related dyad frequencies.

Discussion and conclusions

The results of this study show that temperature has no unidirectional effect on dyad formation in the genotypes studied. This statement is in accordance with the findings of Veilleux and Lauer (1981) and contradicts the conclusion of McHale (1983) who stated that temperature is a major factor under other environmental variables. However, the genotypes designated 'high', used in this study reacted in a similar way to temperature like the high 2n pollen producer with variable expression, used by McHale (1983). The lack of agreement between the dyad frequency at the sporad stage and the frequency of fused spindles at the M_2 might be due to the highly asynchronized course of the meiosis which is often found in potato (Ramanna, 1974). Also in this study different meiotic stages within a single anther were observed in the six genotypes. So it is not excluded that in our material only a single restitution mechanism was responsible for 2n gamete formation at microsporogenesis.

Anomalous cytokinesis can also result in triad and dyad formation. However, it is difficult to prove with light microscopical research which of the new cell walls, the reductional or the equational one, has failed to form (Ramanna, 1974).

The consequence of the possibility that more than one mechanism of nuclear restitution are operative in a given diploid clone, would be desastrous for the reliability of a half tetrad analysis in potato.

Two hypotheses have been put forward to explain the gene regulation of dyad formation in microsporogenesis in potato. Veilleux et al. (1982) suggested an extreme sensitivity of 2n gamete formatin to microenvironmental conditions. They supposed that an extracellular message penetrated into the meiocytes which was thought to be responsible for spindle orientation at the second meiotic division. Consistent expression of the triat may be due to an ability to maintain production of the extracellular message needed for formation of 2n gametes irrespective environmental conditions. McHale (1983) has put forward the hypothesis of homozygosity for a defective allele causing delay in spindle formation or function. This effect may be more pronounced at low temperature.

Although the results of this study indicate a genotype-specific effect of temperature, no evidence could be gained in favour of any of the above-mentioned hypotheses.

Acknowledgement
I thank Ir. P. van Leeuwen, Ms Jacqueline Buurman and Mrs Greet Kuiper for technical assistance and the Department of Horticulture, Agriculture University, Wageningen, the Netherlands for providing the growth chambers.

References
Jacobsen, E., 1976. Cytological studies on diplandroid production in a dihaploid potato clone and its correlation with seed set in 4x x 2x crosses. Z. Pflanzenzüchtg.: 10-15.
Jacobsen, E., 1980. Increase of diplandroid formation and seed set in 4x x 2x crosses in potatoes by genetical manipulation of dihaploids and some theoretical consequences. Z. Pflanzenzüchtg. 85: 110-121.
McHale, N.A., 1983. Environmental induction of high frequency 2n pollen formation in diploid Solanum. Can. J. Genet. Cytol. 25: 609-615.
Mok, D.W.S. & S.J. Peloquin, 1975a. Three mechanisms of 2n pollen formation in diploid potatoes. Can. J. Genet. Cytol. 17: 217-225.
Mok, D.W.S. & S.J. Peloquin, 1975b. The inheritance of three mechanisms of diplandroid (2n pollen) formation in diploid potatoes. Heredity 35: 295-302.
Quinn, A.A., D.W.S. Mok & S.J. Peloquin, 1974. Distribution and significance of diplandroids among the diploid Solanums. Amer. Potato J. 51: 16-21.
Ramanna, M.S., 1974. The origin of unreduced microspores due to aberrant cytokinesis in the meiocytes of potato and its genetic significance. Euphytica 23: 20-30.
Ramanna, M.S., 1979. A re-examination of the mechanisms of 2n gamete formation in potato and its implications for breeding. Euphytica 28: 537-561.
Schroeder, S.H. & S.J. Peloquin, 1983. Seed set in 4x x 2x crosses as related to 2n pollen frequency. Amer. Potato J. 60: 527-536.
Snow, R., 1963. Alcoholic-hydrochloric acid carmine as a stain for chromosomes in squash preparations. Stain Technol. 38: 9-13.

Sokal, R.R. & F.J. Rohlf, 1969. Biometry, the principles and practice of statistics in biological research. W.H. Freeman and Company San Francisco, p. 607-610.

Veilleux, R.E. & F.I. Lauer, 1981. Variation for 2n pollen production in clones of Solanum phureja Juz. and Buk. Theor. Appl. Genet. 59: 95-100.

Veilleux, R.E., N.A. McHale & F.I. Lauer, 1982. 2n gametes in diploid Solanum: Frequency and types of spindle abnormalities. Can. J. Genet. Cytol. 24: 301-314.

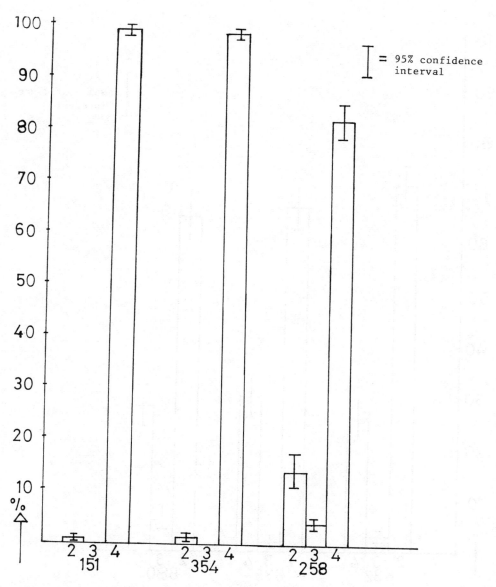

Figure 1. Mean frequency of dyads (2), triads (3) and
tetrads (4) over 4 temperatures for the genotypes 151,
357 and 258.

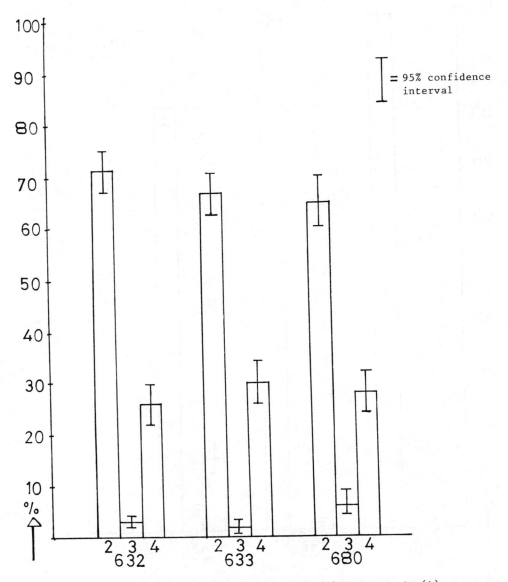

Figure 2. Mean frequency of dyads (2), triads (3) and tetrads (4) over 4 temperatures for the genotypes 632, 633 and 680.

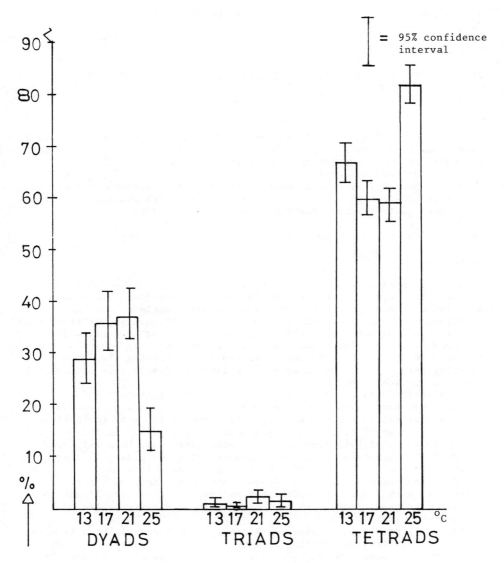

Figure 3. Mean frequency of dyads, triads and tetrads over
6 genotypes, grown at 4 different temperatures.

SEARCH FOR IMPROVED STRATEGY IN THE DEVELOPMENT OF PARENTAL LINES FOR
POTATO BREEDING IN POLAND

K.M. Swiezyński

Institute for Potato Research, Młochów, 05-832 Rozalin, Poland

Summary

Development of parental lines for potato breeding started in Poland 25
years ago. It is now a large scale activity. Its important element is
research done with the intention to improve breeding methods and to
improve the breeding strategy.

Due to parental line development, Polish potato breeders have now a
chance to obtain highly productive new cultivars with multiple resistance
to viruses, late blight, nematodes and storage diseases.

Keywords: potato, breeding, methods.

Introduction

Many old potato cultivars are still widely grown in several countries
with advanced agriculture (Howard, 1980). If we consider, how much
breeding work was done in the meantime, we must conclude that the
efficiency of potato breeding is low. Trying to improve it, we started in
1959 to look for parental combinations producing good progenies. This
appeared to be very laborious. Therefore we decided that creation of
better progenitors would be more reasonable than testing the value of
existing ones. In this way started in Poland the development of parental
lines for potato breeding.

This activity expanded considerably when the Institute for Potato
Research was founded in 1966. The advances of parental line breeding are
being summarized on conferences organized every fifth year (Roguski et
al., 1971, Kapsa et al., 1977, Swiezyński et al., 1984). In the period
1967-1985 in total 1342 samples of parental lines have been delivered to
Polish potato breeders. From these parental lines originate at present in
Poland about 30% of all breeding clones and 3 new cultivars.

We define parental lines as clones, outstanding in some important
characters which, when crossed with a suitable partner, are likely to
produce a progeny, from which new potato cultivars may be selected
(Swiezyński, 1971).

In the first period we made occasionally mistakes, supplying breeders
with parental lines outstanding in some important characters but too poor
in other ones. Some time elapsed before we perceived that parental lines
must be "nearly cultivars".

Objectives in present day parental line breeding

All new potato cultivars in Poland have to be: 1) high yielding, 2)
with acceptable tuber appearance and quality and not too late maturity,
3) at least moderately resistant to potato leafroll virus (PLRV) and
potato virus Y (PVY), 4) wart resistant. The same requirements, slightly
less stringent, must meet parental lines supplied to breeders. In
addition, parental lines produced at present, are outstanding in one or
more of the following characters:
1. High dry matter in the tubers combined with high dry matter yield

2. Adaptation to light, water deficient soils
3. Resistance to viruses.
4. Resistance to <u>Phytophthora infestans</u> in the foliage
5. Resistance to storage diseases
6. Resistance to nematodes.

Most of the listed characters are very complex. To reach advances in some of them a sophisticated breeding program is often needed, which may require several steps of crossing and selection.

The work is done both at the tetraploid and at the diploid level. It is associated with necessary research work.

Development of parental lines at the tetraploid level does not differ much from traditional breeding work. The main difference results from the fact that its objective is not to obtain a new cultivar, but to obtain a better progenitor. Therefore less attention is being paid to full evaluation of individual clones and less care is taken to keep them free from tuber transmissible pathogenes. On the other hand more attention is paid to obtain as soon as possible new favorable combinations of desired characters.

The work at the diploid level is different. It is regarded as long term activity expected to have important bearing on advances of the whole parental line breeding. It is being presented in a separate lecture.

The general scheme of parental line breeding at the tetraploid level.

Usually 4-5 years are regarded sufficient to identify clones deserving using them as parents in own work or to be delivered to breeders as parental lines. The following procedure is typical:
1. Young first year seedlings are being inoculated with pathogenes to reject
   those susceptible to potato virus X (PVX), PVY and P. infestans. Sometimes they are being inoculated also with potato virus S (PVS) or potato virus M (PVM).
   In general all first year seedlings are being grown in the field to obtain a tuber yield easier to evaluate and to propagate.
2. In the first tuber progeny usually 12-30 plants are grown from each clone. The tubers are presprouted and the large ones are cut into pieces, before planting. In this way a sufficient tuber yield is being obtained in the fall to start evaluation for various resistances and to begin testing next year in field trials.
3. The second tuber progeny. The clones are being tested for resistance to pathogenes and they are being evaluated in replicated field trials, using 4-8 hill plots.
4. The third tuber progeny. The clones are being retested for resistance to pathogenes and they are being evaluated in replicated field trials in 2 different locations, using 8-hill plots. The most successful ones are being used next year as parents. They may be also delivered to breeders at the same time or after an additional year of trials. If it is needed to obtain quickly the next generation, true seeds are being produced inwinter time.

Identification of clones with satisfactory yielding ability.

Because of considerable interactions: clones x years and clones x locations, to evaluate the clones, usually means of at least two years of trials are being utilized, advanced clones being evaluated in 2 locations. As chance variation is likely to be large in trials with so small plots, attention is being paid not only to yields of individual clones but also to mean yields of all evaluated clones. We succeeded to reach such a level of yielding ability that usually the mean yield of all tested clones exceeds the yield of cultivars used as standard.

To avoid selection of clones maturing too late, a large proportion of field trials is being harvested in the first half of August (14 weeks after planting with presprouted tubers). This has additional advantages. At that time dry matter content in the tubers is less variable than at autumn harvest (Swiezyński et al. 1977).

Production of clones with multiple resistance.

Most parental lines, delivered to breeders in the last years were high yielding and carried genes for resistance to viruses, late blight or nematodes. Some of them carried the genes Ns and Rm, for resistance to PVS and PVM respectively, which are not yet present in Polish cultivars. In a part of matings performed in 1985 it is already being attempted to combine resistance to most important viruses (PLRV, PVY, PVM) with resistance to P. infestans (in leaves and tubers) and resistance to nematodes (pathotype Ro 1).

Research associated with parental line breeding.

Most research is done in the following 4 areas:
1. Search for new sources of genetic variation. We found interesting sources of high dry matter content (Sawicka, 1984), of resistance to P. infestans (Zarzycka et al., 1984) and of resistance to PVM (Dziewońska, Ostrowska, 1978).
2. Attempts to understand better the pathogenes, the resistance to which is desired. Examples of this type of work is research on PVM (Chrzanowska, 1976, Kowalska, 1981) and research on P. infestans (Sujkowski, 1983).
3. Inheritance of desirable characters in the potato. Some work on resistance to P. infestans and viruses has already been done (Swiezyński et al., 1974, 1981, Butkiewicz, 1978. In progress is further research on inheritance of resistance to these pathogenes as well as on inheritance of high dry matter content, some elements of table quality and of the ability to produce unreduced gamets.
4. Search for improved selection methods. Much work was done specially on genetic and environmental variation in yielding ability, elements of quality and resistance to various pathogenes. Most of the results have been presented in the above cited proceeding of the conferences organized every fifth year and in recent reports (Swieżyński, 1983, 1984).
This research is useful not only for parental line breeding but also for general improvement in potato breeding methods.

Prospects for the future.

We develop parental lines for potato breeding since 25 years. In this period we supplied Polish potato breeders with parental lines outstanding in some quality characters and multiple resistance to some pathogenes. There are good prospects to supply them with still better combinations of characters in the coming years.

This is not enough. There exist sufficient sources of genetic variation to improve potato cultivars in many quality characters and in resistance to numerous pathogenes. As example, the resistance to viruses

140

and Phytophthora infestans, the two most important pathogenes in Poland, may be probably raised considerably, what would result in the reduction of their economic importance.

The breeders will have a much better chance to reach these objectives, if they receive parental lines in the progeny of which desired recombinants will be more frequent than they are now. Three important old papers demonstrate the problem. Simmonds (1969) calculated the limitations to selection for many characters resulting from probability laws. Toxopeus (1953) demonstrated that if parents multiplex for desired genes are utilized, the chance to find recombinants carrying the genes is highly increasing. Hougas en Peloquin (1958) have shown that such parents may be obtained with the use of diploids.

We hope to develop in future parental lines multiplex for several desired genes. Recent publications (MacKay, 1984, Pfeffer, Steinbach, 1982) indicate, that similar attempts are being made in other centers.

References

Butkiewicz, H., 1978. Nietolerancja w stosunku do wirusa liściozwoju występujaca u roślin ziemniaka. Ziemniak. Bonin: 5-37.

Chrzanowska, M., 1976. Variation in symptoms caused by potato virus M on the potato variety Uran. Potato Res. 19: 141-146.

Dziewońska, M.A., K. Ostrowska, 1978. Resistance to potato virus M in certain wild potato species. Potato Res. 21: 129-131.

Hougas, R.W., S.J. Peloquin, 1958. The potential of potato haploids in breeding and genetic research. Am. Potato J. 35: 701-707.

Howard, H.W., 1978. The production of new varieties. Ed. P.M. Harris. The Potato Crop. Chapman and Hall. London: 607-646.

Kapsa, E., K.M. Swieżyński, K. Jastrzebski, 1977. In: Genetyka i Hodowla Ziemniaka. Zesz.Probl.Post.Nauk Rol. 191. PWN Warszawa.

Kowalska, A., 1981. Zmienność wirusów M i S ziemniaka. Instytut Ziemniaka, Bonin.

MacKay, G.R., 1984. Potato breeding. Scott.Crop. Res. Inst. Ann. Report 1983: 62-79.

Pfeffer, Ch., P. Steinbach, 1982. Die Zuechtung von Kartoffelngenotypen mit multiplex Resistenz gegen Globodera rostochiensis. Arch. Zuechtungsforsch. 12: 287-295.

Roguski, K., K. Swieżyński, E. Sawicka, 1971. (eds): Problemy Hodowli Ziemniaka. Zesz.Probl.Post.Nauk Rol. 118. PWN Warszawa.

Sawicka, E.J., 1984. Synteza ziemniaków 24-chromosomowych o wysokiej zawartości skrobi. Zesz.Probl.Post.Nauk Rol. 191: 39-52.

Simmonds, N.W., 1969. Prospects of potato improvement. Scott. Pl. Breed. Sta. 48th Ann. Rep. 1968-69. Pentlandfield: 18-37.

Sujkowski, L., 1983. Zmienność właściwości pasozytniczych grzyba Phytophthora infestans w stosunku do ziemniaka w cyklu rocznym. Agricultural University. Warszawa. Diss.

Swieżyński, K.M., 1971. Ogólne zagadnienia syntezy materialów wyjściowych. Zesz. Probl.Post.Nauk Rol. 118: 9-26.

Swieżyński, K.M. 1983. Parental line breeding in potatoes. Genetika (Yug.) 15: 243-256.

Swieżyński, K.M., 1984. Early generation selection methods used in Polish potato breeding. Am. Potato J. 61 385-394.

Swieżyński, K.M., M.A. Dziewońska, K. Ostrowska, 1981. Inheritance of resistance to potato virus M found in Solanum gourlayi Haw. Genetica Polonica. 22: 1-8.

Swieżyński, k.M., B. Kocyk, E. Kuźmińska, R. Wójcik, 1977. Evaluation of tuber yield starch content in the tubers and some other characters in potato breeding. Genetica Polonica 18: 1-13.

Swieżyński, K.M., J. Pietkiewicz, M.T. Sieczka, 1974. Inheritance of hypersensitivity to Phytophthora infestans and of resistance to viruses in the potato. Genetica Polonica 15: 295-304.

Swieżyński, K.M., E. Sawicka, B. Czech, 1984. (eds) Problemy Hodowli Ziemniaka. Zesz.Probl.Post.Nauk Rol. 273. PWN Warszawa.

Toxopeus, J.H., 1953. On the significance of multiplex parental material in breeding for resistance to some diseases in the potato. Euphytica. 2: 139-146.

Zarzycka, H., E.J. Sawicka, M. Osiecka, L. Sujkowski, 1984. Synteza ziemniaków 24-chromosomowych odpornych na zaraze ziemniaka. Zesz.Probl. Post.Nauk Rol. 191: 53-65.

MONOHAPLOIDS IN POTATO: PSEUDOGAMIC INDUCTION AND POTENTIAL USE IN
GENETIC AND BREEDING RESEARCH

B.A. Uijtewaal and J.G.Th. Hermsen

Department of Plant Breeding (IvP), Agricultural University, Wageningen,
The Netherlands

Introduction

Classical potato breeding is being carried out predominantly by intercrossing preselected tetraploid parental lines which generates genetic
variation, followed by stepwise selection of the best genotypes in extensive field tests over many years in different environments. Once a
cultivar has been obtained, large scale efforts by skilled people are
needed for a healthy maintenance and propagation.

The rapid development of new techniques for cell and tissue culture and
in molecular biology offer new opportunities for generating new variation
by somatic hybridization, for shortening the selection procedures by rapid
in vitro multiplication of advanced genotypes and – on the long term and
still speculative – by improving existing cultivars in specific traits by
means of transformation procedures.

Somatic hybridization combines the intact idiotypes of two somatic cells
which – after regeneration – leads to plants having the total chromosome
complement of both parents in a mixture of the parental cytoplasms, if
elimination of chromosomes or cytoplasm of one parent does not occur. As
in potato the tetraploid level is generally considered to be optimal for
varietal performance, intraspecific somatic hybridization should be
preceeded by breeding at the diploid level.

In addition to the characteristic mentioned in the previous paragraph
somatic hybridization is independent of gametes and thus can be carried
out also with non-flowering or fully sterile plants and with plants in
very early growth stages. So monohaploids having only one genome (x=12 in
potato)can directly be hybridized.

Theoretically monohaploids are a powerful tool in genetic research
because both dominant and recessive mutants are directly visible. In
addition they are the only way of producing homozygous potatoes at different ploidy levels.

In this article the production of monohaploids and their potential use
will briefly be discussed.

The production of monohaploids

More than 400 monohaploids from 12 different diploid genotypes have been
produced through pseudogamy using as pollinator Solanum phureja clones
that are homozygous for the dominant marker 'embryo-spot'. These clones
were developed by Hermsen and Verdenius (1973) and have a high dihaploid
inducing ability in autotetraploid potatoes. When used as pollinator onto
diploid potatoes, monohaploids may be obtained from certain genotypes.
However, their frequency is extremely various, dependent of the diploid
genotype and on an average it is such low, that only the embryo marker
enables their detection. Embryo spot marks the hybrid embryos, haploid
embryos being spotless. The highest frequency found up to now has been one
per 1000 seeds (Van Breukelen et al., 1975; Uijtewaal, unpublished data).

143

Monohaploid production has also been attempted using anther culture, but hitherto without success. Other researchers were successful (Sopory et al., 1978; Foroughi-Wehr et al., 1977; Veilleux et al.. 1985), but only with few specific genotypes. Breeding for 'anther culture ability' seems to be feasible (Tanaka, 1985, unpublished).

Assessment of monohaploids

Monohaploid plants were studied for vigour and tuber production. In vitro assessment was carried out for growth vigour, stability of ploidy level, ability to produce callus and to regenerate plants from callus. The quality of different monohaploids at the protoplast level was studied as well. Out of a total number of about 400 monohaploids, 40 were selected for being used in research at the protoplast level.

The use of monohaploids in research at IvP

Monohaploids have potentials but limitations as well. Some aspects of their use in research at IvP will be discussed in this chapter (see scheme in Figure 1).

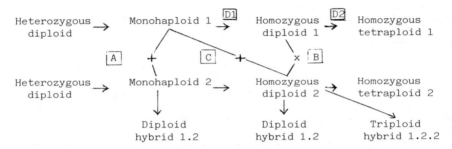

Fig. 1. Scheme for utilizing monohaploids involving both sexual (x) and somatic (+) hybridization.

Explanation of scheme

A in figure 1 represents the fusion of protoplasts from the monohaploids 1 and 2 yielding the diploid somatic hybrid 1.2. B marks sexual hybridization between the doubled monohaploids (= homozygous diploids) 1 and 2 resulting in the diploid hybrid 1.2, which is identical to that under A. Triploid hybrids can be produced in different ways, one of which is fusing a monohaploid with a diploid, as indicated by C in fig. 1. Homozygous diploids and tetraploids can be obtained by identical doubling of a monohaploid and the corresponding homozygous diploid respectively (D1 and D2 in fig. 1) (Hermsen et al., 1981).

Drawbacks related to handling monohaploids

Monohaploids being derived (in two steps) from natural autotetraploid potato tend to return to a higher ploidy level spontaneously (Jacobsen et al., 1983; Karp et al., 1984; these authors, unpubl.).
Flow cytometer determination of ploidy levels in leaf cells from monohaploid plants revealed that 30-50% of these cells are diploid instead of monoploid. This implies that 30-50% of the protoplasts isolated from the leaves are not monoploid anymore. Fusion of such leaf protoplasts without

previous assessment of ploidy levels will yield erratic results.

Fusion of protoplasts from a monohaploid with those from a diploid (Fig. 1C) should theoretically give rise to triploids. In this case, fusion products may be selected on the basis of ploidy level. Such system might be used for research in optimizing the fusing techniques. Another criterion used for selection of hybrid fusion products is based on hybrid vigour of the fusion products. Hybrid vigour may occur in somatic hybrids from parents with a good combining ability for growth in vitro. Our experiments indicated the feasibility of this selection criterion, but no proof is available yet.

Homozygous ploidy series

Several series of homozygous monoploid-diploid-tetraploid clones (D1 and D2 in Fig. 1) have been produced and are being used for research on gene dosage effects and optimum ploidy level in homozygous potato regarding plant vigour.

Selection for vigour of monohaploids implies selection for ideal combinations of genes. Only monohaploids and derivatives offer an opportunity to investigate, whether it is feasible to obtain homozygous potato clones that may either approach the performance of a cultivar or be used as a parent in a single or double cross scheme leading to well yielding and uniform hybrid potatoes from true seeds as an alternative to the use of 2n-FDR gametes for this purpose (cf. the contribution by E. Jongedijk). At present six series of homozygous x, 2x and 4x lines are being investigated.

Concluding remarks

Theoretically monohaploids are ideal material for basic and applied research. However, their instability in vivo and in vitro is a serious drawback. On the other hand, we found differences in degree of instability. Selection for increased stability may be feasible.

The large number of monohaploids produced at IvP, in addition to their diversity in vigour, morphology, tissue culture ability and other traits, warrant more solid data about the potential use of monohaploids in potato research of to-morrow.

References

Breukelen, E.W.M. van, M.S. Ramanna and J.G.Th. Hermsen, 1975. Monohaploids (n=x=12) from autotetraploid Solanum tuberosum (2n=4x=48) through two successive cycles of female parthenogenesis. Euphytica 24:567-574.

Foroughi-Wehr, B., H.M. Wilson, G. Mix and H. Gaul, 1977. Monohaploid plants from anthers of a dihaploid genotype of Solanum tuberosum L. Euphytica 26:361-367.

Hermsen, J.G.Th. and J. Verdenius, 1973. Selection from Solanum tuberosum group Phureja of genotypes combining high-frequency haploid induction with homozygosity for embryo-spot. Euphytica 22:244-259.

Hermsen, J.G.Th., M.S. Ramanna, S. Roest and G.S. Bokelmann, 1981. Chromosome doubling through adventitious shoot formation on in vitro cultivated leafexplants from diploid interspecific potato hybrids. Euphytica 30: 239-246.

Jacobsen, E., M.J. Tempelaar and E.W. Bymolt, 1983. Ploidy levels in leaf callus and regenerated plants of Solanum tuberosum determined by cytophotometric measurements of protoplasts. Theor. Appl. Genet. 65:113-118.

Karp, A., R. Risiott, M.G.K. Jones and S.W.J. Bright, 1984. Chromosome doubling in monohaploid and dihaploid potatoes by regeneration from cultured leafexplants. Plant Cell Tissue Organ Culture 3:363–373.

Sopory, S.K., E. Jacobsen and G. Wenzel, 1978. Production of monohaploid embryoids and plantlets in cultured anthers of <u>Solanum</u> <u>tuberosum</u>. Plant Science Letters 12:47–54.

Veilleux, R.E., J. Booze-Daniels and E. Pehu, 1985. Anther culture of a 2n pollen producing clone of <u>Solanum</u> <u>phureja</u> Juz. & Buk. Can. J. Genet. Cytol. 27:559–564.

PROTOPLAST TECHNOLOGY; SOMATIC HYBRIDIZATION AND CULTURE IN POTATO
BREEDING PROGRAMMES

S. Roest and K.J. Puite

Research Institute ITAL, P.O. Box 48, 6700 AA Wageningen, The Netherlands

Summary

Protoplasts isolated from green shoots of dihaploid S. tuberosum (2n=24)
were electrofused with fluorescein diacetate stained protoplasts from
herbicide-bleached shoots of diploid S. phureja (2n=24). Heterokaryons
could be identified by red autofluorescence of green chloroplasts and
yellow-green FDA fluorescence of bleached protoplasts. They were selected
with a micromanipulator. Using various feeder systems plating of heterokaryons at relatively low densities resulted in the production of
vigorously growing calli and the subsequent regeneration of adventitious
shoots. Cytological analysis of one of the regenerated plantlets revealed
the presence of chromosomes of both fusion partners and the tetraploid
number of chromosomes (48) in most cells.
Keywords: protoplasts, somatic hybridization, electrofusion, potato,
Solanum.

Introduction

Breeding of the potato at the tetraploid level is complicated (laborious
and time-consuming) and may lead to segregation of desirable characters.
Breeding and selection at the diploid level is more efficient and can considerably accelerate the breeding process. Somatic hybridization of two
selected diploid parental lines prevents segregation and in principle
gives rise to a tetraploid genome and a high degree of heterozygosity. In
view of a joint project with the Foundation for Plant Breeding (SVP) at
Wageningen on the possibilities of somatic hybridization for potato
breeding, it has been investigated at the Research Institute ITAL whether
protoplasts of dihaploid S. tuberosum and diploid S. phureja can be isolated, fused, cultured and regenerated to hybrid plants.

Materials and methods

The dihaploid SVP 1 line (SH 77-78-1994) of S. tuberosum and the diploid
SVP 5 line (PH 77-1445-2242) of S. phureja were used and with some modifications the procedure of isolation, culture and regeneration of shoot cul-
ture-derived protoplasts was followed as described for the tetraploid com-
mercial cultivar Bintje (Bokelmann and Roest, 1983; Roest and Bokelmann,
1983).
Green shoots of SVP 1 were grown on MS medium supplemented with 1% sucrose, whereas shoots of SVP 5 were bleached on herbicide SAN 9789 containing MS medium supplemented with 3% sucrose (Uhrig, 1981).
Protoplasts of both lines were isolated and the bleached protoplasts were
stained with fluorescein diacetate (FDA). A 1:1 mixture of the parental
protoplasts (at 10^5 protoplasts per ml) was electrofused in 0.5 M mannitol
in a 1 ml multi-electrode fusion chamber, being open underneath and having
a 3 mm electrode distance (Puite and Roest, 1985; Puite et al., 1985). The
fusion process was followed under a fluorescence microscope and the

formation of heterokaryons was easily identified by red autofluorescence
of the green chloroplasts of SVP 1 and green-yellow FDA fluorescence of
SVP 5. Directly after electrofusion the electrodes were removed and 1 ml
of an adapted culture medium was added, resulting in the normal composition of the medium and a density of 5.10^4 protoplasts per ml. One and
two days after electrofusion 840 heterokaryons were selected with a micromanipulator and each time 120 heterokaryons were transferred into the
center well of Falcon dishes in 0.1 or 0.2 ml of liquid culture medium,
using different feeder systems. Some experiments were carried out with
irradiated and in agarose embedded SVP 1 protoplasts in the center well,
whereas a cell suspension culture of a mutant line of Nicotiana plumbaginifolia (the NA36 line) or culture medium was present in the outer part
of the dish.
Twelve days after electrofusion 0.2 ml of fresh liquid culture medium was
added to the center well and 3 days later the developing cell aggregates
were transferred with a micropipette to 1.5 ml of fresh culture medium and
embedded in agarose 0.2%. Four weeks after electrofusion the cell colonies
were picked up with tweezers and plated on top of a solid growth medium.
Two weeks later the calli were transferred to shoot initiation and subsequently to shoot elongation medium. For plantlet production some shoots
were excised from the calli and subcultured on a medium for root
formation. One of the regenerated plantlets was analysed for Giemsa C-band
pattern of metaphase chromosomes in shoot and root tip cells (Pijnacker
and Ferwerda, 1984).

Results and discussion

Via electrofusion heterokaryon formation frequencies of $6 \pm 1\%$ were
scored. While protoplasts are usually cultured at a density of $5.10^4 - 10^5$
per ml for cell wall formation and subsequent regeneration, these
experiments have shown that 120 heterokaryons can be successfully cultured
in 0.2 or 0.1 ml of culture medium at corresponding densities of 0.6×10^3
and 1.2×10^3 per ml. It is not clear whether regeneration at such relatively low densities has to be ascribed to the preculture during 1 or 2
days at an almost optimal density of 5.10^4 protoplasts per ml, to the
feeder effect of cells and/or protoplasts, or to a possible hybrid vigour
of developing heterokaryons.
No substantial difference in plating efficiency (P.E.) was observed between the different feeder systems. Selection of heterokaryons after 2
days, however, proved to be more efficient (P.E. 6%) than after 1 day
(P.E. $\pm$ 3%), which may be due to mechanical damage because the process of
cell wall formation has not yet completed 1 day after fusion.
Plated heterokaryons have shown the first cell divisions after 5 days and
the developing cell aggregates (Fig. 1) are 0.5-1 and 1-2 mm in diameter,
2 and 4 weeks after electrofusion, respectively. From 840 selected and
cultured heterokaryons 41 putative hybrid calli were obtained (P.E. $\pm$ 5%),
growing much more vigorous on solid growth medium than the calli obtained
from protoplasts of the parental lines, which may indicate hybrid vigour.
So far 10 out of 41 calli, transferred to shoot initiation and shoot
elongation medium, regenerated adventitious shoots and via root formation
of excised shoots the first plantlets were produced. The hybrid nature of
one of the regenerated plantlets was confirmed by Giemsa C-banding of
metaphase chromosomes, demonstrating the presence of chromosomes of S.
tuberosum and S. phureja, and the tetraploid number of chromosomes (48) in
most cells. In order to establish the hybrid nature of other regenerated
plants, further morphological, cytological and isozyme analysis will be
carried out.

148

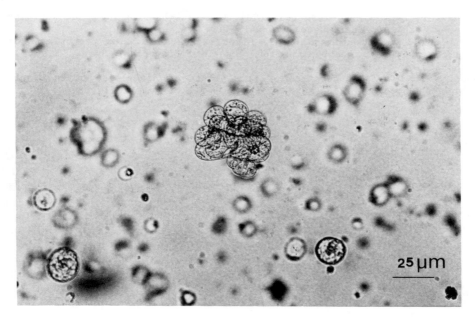

Fig. 1. Cell aggregate which developed from a selected heterokaryon 1 week
after electrofusion and plating on top of in agarose embedded irradiated SVP 1 protoplasts.

Conclusion

It can be concluded that electrofusion has shown to be a relatively simple
and reproducible procedure. Manual selection and culture of heterokaryons
has resulted in the formation of putative hybrid calli and the subsequent
regeneration of the first shoots and plantlets. The hybrid nature of one
of the regenerants was confirmed by cytological observations.

Acknowledgement

The transfer of the heterokaryons with the micromanipulator was carried
out by Mr. P. van Wikselaar and Ms. M. Zaal participated in the feeder
layer experiments. Dr. L.P. Pijnacker and Ms. M.A. Ferwerda (Dept. of
Genetics, University of Groningen) carried out the chromosome banding
technique and analysis.

References

Bokelmann, G.S. & S. Roest, 1983. Plant regeneration from protoplasts of
   potato (Solanum tuberosum cv. Bintje). Z. Pflanzenphysiol. 109 (3):
   259-265.
Puite, K.J. & S. Roest, 1985. Somatic hybridization between two Nicotiana
   plumbaginifolia lines and between Solanum tuberosum and S. phureja using
   electrofusion. Poster Abstract EUCARPIA Symp. "Genetic manipulation in
   plant breeding", W.-Berlin, 8-13 Sept.
Puite, K.J., P. van Wikselaar & H. Verhoeven, 1985. Electrofusion, a
   simple and reproducible technique in somatic hybridization of Nicotiana
   plumbaginifolia mutants. Plant Cell Reports (in press).

Pijnacker, L.P. & M.A. Ferwerda, 1984. Giemsa C-banding of potato chromosomes. Can. J. Genet. Cytol. 26: 415-419.
Roest, S. & G.S. Bokelmann, 1983. Plant regeneration from protoplasts of
different potato genotypes. Proc. 6th Int. Protoplast Symp., Basel, p
282-283.
Uhrig, H., 1981. Regeneration of protoplasts of dihaploid potato plants
bleached by a herbicide (SAN 6706). Mol. Gen. Genet. 181: 403-405.

Posters

Virulence spectra of Globodera pallida Pa-3 isolates and resistance spectra of Pa-3 resistant potato plantgenotypes

L.M.W. Dellaert and J.H. Vinke

Foundation for Agricultural Plant Breeding (SVP), P.O. Box 117, 6700 AC Wageningen, the Netherlands

Introduction

To test resistance to the potato cyst nematode Globodera pallida pathotype Pa-3 various Pa-3 isolates are used for inoculation. The identification of the nematode isolates as Pa-3 is based on the virulence of the Pa-3 isolates on the Pa-2 differential $(VTN)^2$ 62-33-3. Therefore the Pa-3 isolates may be composed of different pathotypes and thus may show differences in virulence spectra on resistant plant genotype. Vice versa also the 'Pa-3' resistant plant genotypes may show differences in resistance spectra.

To test this hypothesis twelve selected potato clones (AM-clones) with resistance to Globodera rostochiensis and G. pallida virulence groups, Ro-1/Ro-4, Ro-2/Ro-3, Pa-2 and Pa-3, and four susceptable standard varieties, namely Saturna, KTT 60-21-19, ODV 22731 and $(VTN)^2$ 62-33-3, were inoculated with eleven different G. pallida Pa-3 isolates in 1982, 1983 and 1984.

Results

Compared to the four susceptible standard varieties the reproduction rate of the Globodera pallida Pa-3 isolates is significant less on the selected resistant clones. The reproduction rate expressed as Pf/Pi (Pf = average final number of cysts; Pi = initial number of cyst) of the resistant AM clones is less than one, whereas the Pf/Pi ratio on the standard varieties varies from three for $(VTN)^2$ 62-33-3 to fifteen for Saturna.

Within the standard varieties and within the group of AM-clones significant main effects of plantgenotype, Pa-3 isolate and year are observed. Also the interactions plantgenotype x Pa-3 isolate, plantgenotype x year and Pa-3 isolate x year are significant.

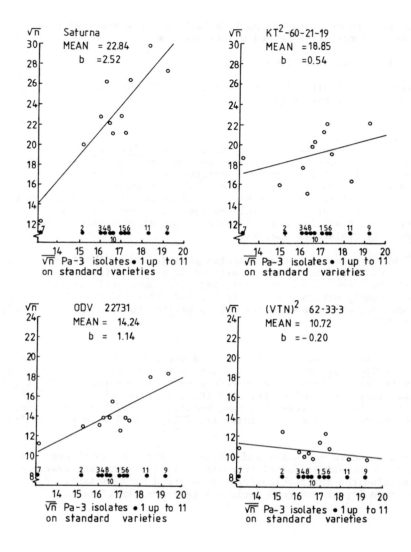

Fig. 1. Regression of the resistant level (Vn) of the standard varieties Saturna, KT[2] 60-21-19, ODV 22731 and (VTN)[2] 62-33-3 on the Pa-3 isolates characterized by the average virulence $\sqrt{n}$ (n=number of cysts) on the standard varieties. Y = the average Vn per Pa-3 isolate.

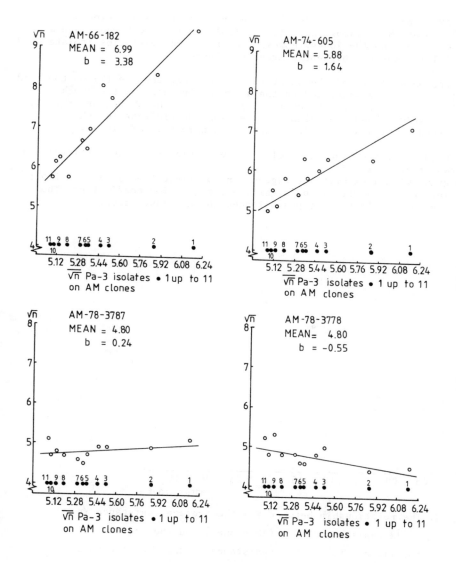

Fig. 2. Regression of the resistant level (Vn) of the plant genotypes
AM-66-182; AM 74-605; AM 78-3787 and AM 3778 on the
Pa-3 isolates characterized by the average virulence $\overline{Vn}$
(n = number of cysts) on the resistant clones. Y = average Vn
per Pa-3 isolate.

The AM-clones 78-3778 and 78-3787 are the most resistant clones. The
Pa-3 isolate No 1 (Rookmaker) is the isolate with the highest average
virulence on the AM-clones. However, on the check varieties isolate No 9
(coll 1077) has the highest average virulence.

Regression analysis of the plant resistance per genotype (Y) on the Pa-3
isolates (X), characterized by the average virulence on the check
varieties (Fig. 1) and the average virulence on the AM-clones (Fig. 2)
respectively, shows significant regression coefficients for the
cultivars Saturna and ODV 22731 and for a number of AM clones.

Based on the regression coefficient, the AM clones can be distinguished
in two groups (Fig. 3).
One group with a significant regression coefficient varying from 1.5 to
3.5 and one group with a non significant regression coefficient. The
most resistant plant genotypes belonged to this last group.

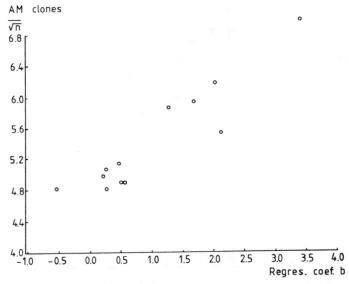

Fig. 3. The relationship between Y = the average resistance $\sqrt{n}$
of 12 resistant clones (AM-clones) and the regression
coefficient (b) of the resistance of the twelve clones
on the Pa-3 isolates.

Conclusion
"Pa-3 isolates" show differences in virulence spectra and "Pa-3
resistant" clones show differences in resistance spectra. Therefore the
isolates differ in the frequency of corresponding virulence genes, and
the AM-clones differ in the frequency of corresponding resistance genes.

To select the AM-clones with the highest resistant level selection
should be based on resistant tests with different isolates.

THE GERMAN-NETHERLANDS POTATO COLLECTION

R. Hoekstra

Institut für Pflanzenbau und Pflanzenzüchtung der FAL, Brunswick, Federal
Republic of Germany, in cooperation with the Foundation for Plantbreeding
(SVP) and the Gene Bank Netherlands (GBN), Wageningen, Netherlands

Summary

The German-Netherlands Potato Collection contains about 2900 accessions
of wild species and primitive forms of the potato. The available germplasm
is extremely varied and is a valuable source of genetic variability for
potato breeding. A systematic evaluation programme tests the collection
for resistance to pests and diseases, which are important breeding
properties. The information in the computer supported data base is quickly
available and easy of access. A new publication of all the evaluation data
is foreseen.
Keywords: potato germplasm, wild species, primitive forms, maintenance,
           evaluation, documentation.

Introduction

In 1974 a cooperation between the Federal Republic of Germany and the
Kingdom of the Netherlands started in the field of the genetic resources
of the potato at the "Institut für Pflanzenbau und Pflanzenzüchtung der
FAL. The netherlands partner in this cooperation is the Foundation for
Plantbreeding (SVP). The objectives of the project are:
- the establishment of a germplasm collection with wild species and
  primitive forms of the potato
- maintenance and storage of the collection
- coordination of an evaluation programme on important properties
- documentation of the collection
- publication of the data and distribution of the germplasm
- carrying out scientific research on the collection.
A broad collection has been established with important tuber-bearing
Solanum species as Solanum chacoense, S. demissum, S. microdontum,
S. sparsipilum, S. spegazzinii, S. vernei and the primitive form S. tuberosum subsp. andigena represented with many accessions.

Maintenance

The collection is stored in the form of true seed. It includes 2100
accessions of 90 wild tuber-bearing Solanum species and 750 accessions of
5 primitive forms. Rejuvenation is carried out by hand pollination in the
glasshouse and by growing self-incompatible species under spatial isolation
in cereals in the field. Due to quarantine regulations, the cultivation in
the field is temporarely prohibited in Germany, until possibilities to
test the plants are created. Starting in 1984 the SVP/GBN participates
in the rejuvenation of the collection by growing 20 accessions in the field
at Wageningen, after the dutch Plant Health Service (PD) tested the plants.
Each year about 120 accessions are rejuvenated successfully. The long-term
storage at Brunswick (-10°C, 5-7 % moisture content) makes a storage period
of 20-30 years attainable, without significant loss of viability.

Germination tests are carried out regularly to test the viability of the seeds. From about 50 % of the accessions of the collection seeds are available on request.

Systematic evaluation programme

For the utilization of the genetic resources of the potato the systematic evaluation of the collection is of great importance. Specialists in both cooperating countries are testing the potato germplasm pool on properties that have a high priority in the potato breeding programmes. Table 1 illustrates the current screenings for resistance to pests and diseases, carried out by the specialists.

Table 1. Systematic evaluation programme.

Caracter	Research worker
Blackleg	Dr. Munzert (BLPB, Freising)
Globodera pallida (PA$_2$)	Dr. Rumpenhorst (BBA,Münster)
Globodera pallida (PA$_3$)	Dr. Dellaert (SVP,Wageningen)
Late blight (laboratory test)	Dr. Schöber (BBA,Braunschweig)
Late blight (field test)	Ir. Beekman (SVP,Wageningen)
Wart (pathotype 2 & 6)	Dr. Langerfeld (BBA,Braunschweig)

BBA = Biologische Bundesanstalt; BLPB = Bayerische Landesanstalt für Bodenkultur und Pflanzenbau; SVP = Stichting voor Plantenveredeling

The evaluation data have been discussed in several publications (van Soest, 1983; van Soest et al., 1983; 1984). Table 2 reviews the number of evaluation data for each evaluated character.

Table 2. Survey of the evaluation data (August 1985).

Character	Number of accessions tested	Number of species tested
Globodera rostochiensis	1081	50
Globodera pallida	1087	59
Blackleg	350	56
Late blight	2255	87
Wart	520	48
Dry rot	487	37
Common scab	22	1
Potato Virus M	237	12
Potato Virus X	238	50
Potato Virus Y	243	55
Vitamin C	75	2

Documentation and distribution

The passport and evaluation data from the computer supported data base can be printed in almost any requested form. The information is quickly available and easy of access. After the publication of the complete evaluation data (van Soest & Seidewitz, 1981) more than 2000 new data have been added, and a new publication is foreseen. Excluding the samples used for the evaluation work, 1356 samples were distributed from 1981 till 1985. About a third part of these samples were directly used for potato breeding purposes.

Research on blooming and seed set

Regarding the maintenance of the collection, the seed production of some species like S. oplocense and S. tuberosum subsp. andigena, often is dissatissfactory. Methods are to be determinated for the improvement of blooming and seed set, to make the rejuvenation of these species more efficient. The first results from the experiment with plant growth substances indicate the possibility to improve fruit set by spraying 100 ppm p–CPA or 2,4 D on the flower, one day after pollination.

References

Soest, L.J.M. van, 1983. Evaluation and Distribution of important properties in the German-Netherlands Potato Collection. Potato Res. 26:109-121.

Soest, L.J.M. van, H.J. Rumpenhorst & C.A. Huijsman, 1983. Potato cyst-nematode resistance in tuber-bearing Solanum species and its geographical distribution. Euphytica 32:65-74.

Soest, L.J.M. van, B. Schöber & M.F. Tazelaar, 1984. Resistance to Phytophthora infestans in tuber-bearing species of Solanum and its geographical distribution. Potato Res. 27:393-411.

Soest, L.J.M. van & L. Seidewitz, 1981. Evaluation data on tuber-bearing Solanum species. Institut für Pflanzenbau und Pflanzenzüchtung der FAL - Stichting voor Plantenveredeling (SVP), 165pp.

THE CROSSABILITY OF <u>SOLANUM TUBEROSUM</u> WITH TWO WILD SPECIES, SERIES LONGIPEDICELLATA, RESISTENT TO LATE BLIGHT.

Louis J.M. van Soest[1]

Centre for Genetic Resources Netherlands (CGN)-Foundation for Agricultural Plant Breeding (SVP), Wageningen, the Netherlands

Summary

Attemps were made to transfer genes of two tuber-bearing <u>Solanum</u> species, <u>S.hjertingii</u> Hawkes and <u>S.fendleri</u> A.Gray, into cultivated potatoes. Both species combine several interesting characters and are therefore considered to have high potential value for potato breeding. The results of interspecific crosses between the two species on the one hand and cultivated potatoes, dihaploids and wild species on the other are presented. The crossability of <u>S.hjertingii</u> and <u>S.fendleri</u> with several potato cultivars was very difficult. However, some three-way hybrids were obtained when the cv. Olympia as female parent was fertilized with a pollen mixture of the cross <u>S.fendleri</u> x <u>S.hjertingii</u>. These three-way hybrids were further backcrossed with cultivated tetraploids, but relatively low seed set was obtained. Reciprocal hybridization between the two allotetraploids on the one hand and four dihaploids and three wild species of the Series Tuberosa on the other, yielded in several sterile triploids. The late blight resistance of several hybrids was tested and high levels of resistance were found. The three-way hybrids were repeatedly tested during two successive years and the average level of resistance of this material was high.
Finally, possible schemes to transfer valuable genes of Longipedicellata species into the cultivated tetraploid potato are discussed.

Introduction

<u>S.hjertingii</u> and <u>S.fendleri</u> are allotetraploid species belonging to the Series Longipedicellata (LON) and are mainly present in northern Mexico. These species are well known for their resistance to <u>Phytophthora infestans</u> (Niederhauser and Mills, 1953; van Soest et al, 1984; Toxopeus, 1964 and Umaerus and Stalhammar, 1969). Besides high levels of late blight resistance, <u>S.hertingii</u> and <u>S.fendleri</u> have several other interesting properties;
. high resistance to potato aphids (Radcliffe et al, 1981),
. fairly high levels of resistance to <u>Erwinia carotovora</u> var. <u>atroseptica</u> (Munzert, personal communication),
. prolonged tuber-seed dormancy (personal observation),
. tubers of reasonable shape without deep eyes (personal observation),
. lack of enzymic browning in <u>S.hjertingii</u> only (Firbas, 1961), which can be transferred into hybrids progenies (Woodwards and Jackson, 1985).
The allotetraploid LON species are extremely difficult to cross directly with <u>S.tuberosum</u> (Radcliffe et al, 1981; van Soest, 1983 and Woodwards and Jackson, 1985). Only in Denmark, Foldo (personal communication) obtained direct hybrids between <u>S.tuberosum</u> and <u>S.hjertingii</u>.

1. Research conducted between 1980-1983 at the German-Netherlands Potato Department of the Gene Bank in the Federal Research Centre for Agriculture Braunschweig-Volkenrode (FAL).

There is cytological evidence that the LON species are allotetraploids (Marks, 1958 and Hawkes, 1966) and the following genome symbols have been postulated: A_4 B. The A_4 genome is very similar to the A_1 genome found in many diploid wild species and this explains the occurrence of sterile triploids (A_1 A_4 B) between LON species and some diploid tuber-bearing wild species. In view of the reported problems of direct crosses with S.tuberosum, it seems that hybridization may be only successful when bridging species or other forms of genetic manipulation are employed. This paper describes the crossing experiments with the allotetraploid species S.hertingii and S.fendleri and the non-race specific late blight resistance of some progenies.

Material and Methods.

The material used in this study is presented in table 1. Initially 10 potato varieties were utilized, but later only Olympia and Apatit. (Table 1).

Table 1. Material utilized in the crosses and their late blight resistance

Varieties/Species	Abbreviation	Accession number	2n	late blight resistance	
				I	II
cv. Olympia	OLY	-	48	7/8 (4)	3
cv. Apatit[2]	APA	-	48	2 (1)	3
BGRC 28987	-	BGRC 28987	48	9 (4)	-
S.hjertingii	hje	BGRC 8091	48	1 (7)	2
S.fendleri	fen (1)	BGRC 8088	48	2 (4)	6
	fen (2)	BGRC 8090	48	1 (7)	7
S.microdontum	mcd	BGRC 18503	24	3 (3)	2
S.berthaultii	ber	BGRC 18548	24	2 (2)	1
Dihaploiden[3]	DH	-	24	9 (9)	-

1. I= laboratory test (van Soest et al, 1984). Number in parenthesis denotes total number of tests. II= field test (van Soest et al, 1984)
2. breeding population from Denmark, with S.hjertingii in its ancestry (received from Dr.Foldo)
3. four different fertile dihaploids

Crosses were carried out in the convential way and pollen fertility was estimated using lactophenol acid fuchsin. The late blight resistance of the basis material is also presented in table 1. The resistance was predominantly tested by means of a laboratory test but some material was screened with a field test. Both methods are described by van Soest et al (1984). Inoculation is conducted with mixtures of complex races. These mixtures should overcome most modern varieties which contain the appropiate compatible R genes; thus within evident limit failures could represent examples of non-specific resistance. The degree of resistance is determined by means of 1 to 9 scale (1=very low susceptible; 9= very high susceptible), and based on formulea for the sporulation index (SI) and the weighted mean (WM) for the laboratory- and field test respectively (van Soest et al, 1984).

Results

Crossing experiments

The following conclusion can be drawn
- Direct crosses between hje or fen and potato varieties (4x) remain extremely difficult. Only sterile cripples were obtained from the crosses OLY x hje and hje x OLY.
- Reciprocal crosses between hje, fen and hybrids of both species on the one hand and four dihaploids and two diploids wild species (Series Tuberosa) on the other all resulted in vigorous, completely sterile, allotriploids.
- 18 sometimes very vigorous interspecific hybrids, derived from 9 progenies (berries), were obtained from the cross OLY x (fen (1) x hje), whereas the other three-way crosses yielded mainly cripples and plants of reduced vigour (table 2)

Table 2. Results of three-way crosses in 1981.

| | | | | | Reciprocal cross | | | |
Cross	n	b	s	vs	n	b	s	vs
(fen(1) x hje) x OLY	62	1	4	0	35	26	42	18**
(fen(1) x hje) x APA	61	4	4	0	7	-	-	-
(feb(2) x hje) x OLY	31	0	-	-	3	-	-	-
(fen(2) x hje) x APA	19	0	-	-	2	-	-	-
(hje x fen(2)) x OLY	102	5	11	5*	7	2	0	-
(hje x fen(2)) x APA	59	3	3	1	21	-	-	-

n=number of pollinated flowers, b=berries, s=seeds.
vs=viable seeds.
 * 4 plants of reduced vigour, showing abnormal growth and no flowers (disharmony of parental genomes).
 **only a few plants showed abnormal growth and did not flower.

The hybrids OLY x (fen(1) x hje) could be seperated into plants resembling the cross (fen x hje), intermediates and plants morphologically more similar to Olympia. 16 different three-way hybrids were further crossed as pistillate parents with Olympia, Apatit and BGRC 28987. Only four plants could be obtained from 392 crosses after pollination with BGRC 28987. Two of these plants were cripples whereas the other produced vigorous plants. Finally, one of these two plants was crossable with Olympia and BGRC 28987, producing several hundreds of seed in 1983.

Late blight resistance

The non-race specific late blight resistance of progenies including hje and fen genes is presented in table 3. The resistance levels of the allotriploids between the LON species and ber and mcd were all very high (scale 1-2). The levels of the three-way crosses with the highly susceptible dihaploids were rather variable (scales 1-6

Table 3. Non-race specific late blight resistance of progenies of several
         interspecific crosses with hje and fen.

Cross	2n	Late blight resistance[1]	
		I	II
fen(1) x hje	48	1 (5)	3
fen(2) x hje	48	5/6 (3)	5
hje x fen(1)	48	1 (3)	3
fen(1) x ber	36	1/2 (6)	1
hje x ber	36	2 (4)	1
fen(1) x med	36	1 (2)	-
DH* x (fen(1) x hje)	36	6 (4)	-
DH* x (hje x fen(1))	36	1 (1)	-
(fen(1) x hje) x DH*	36	1 (2)	-
(fen(2) x hje) x DH*	36	5 (2)	-
(hje x fen(1) x DH*	36	1 (2)	-
OLY x (fen(1) x hje)[2]	72?	2/3 (22)	-
OLY x (fen(1) x hje) xBGRC 28987[2]	48?	5/6 (5)	-

* in most cases two different dihaploids
1. I=laboratory test. Number in parenthesis denotes total number of tests.
   II=field test (only one test)
2. plants were vegetatively propagated to make more tests possible.

   The nine OLY x (fen(1) x hje) progenies show variable levels of
resistance (table 4). In average the resistance is high (scale 2/3) but
three progenies showed medium levels of susceptibility (scale 5/6). The
level of P.infestans resistance of the hybrid (OLY x (fen(1) x hje)) x
BGRC 28987 was variable, on average scale 5/6. Material of this hybrid was
vegetatively propagated and five successive tests for resistance showed
scales from 3 to 9. BGRC 28987 itself is highly susceptible to late blight
(scale 9).

Table 4. Late blight resistance of the nine OLY x (fen(1) x hje)
         progenies. (results of the laboratory test only)

Progeny	No. of genotypes	Scale	Number of test
1	3	3	3
2	2	5	2
3	3	1	3
4	2	6	3
5	2	5	2
6	1*	1	1
7	1*	1	1
8	1*	2	2
9	3	1	5

*vegetatively propagated (5 plants)
Scale 1=very low susceptible, 9=very high susceptible

Discussion

From the experiments it became clear that direct crosses between LON species and S.tuberosum are extremely difficult. Fig 1 presents three possible schemes to obtain hybrids. It seems that hybridization can be successful when first allohexaploids ($A_1 A_1 A_4 A_4$ BB) are produced. In this experiment hybrids were obtained according to the first scheme (unreduced gametes). There is some evidence that hybrids of two LON species are easier to hybridize with S.tuberosum than the single species. This agrees with the findings of Stelzner (1943) with the (allo)tetraploid S.acaule. It is suggested that such interspecific hybrids produce more unreduced gametes.

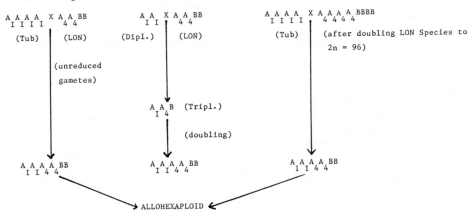

Fig 1: Schemes for transferring genes from LON species into cultivated potatoes

LON   = Longipedicellata ($A_4 A_4$ BB)

Tub   = S.tuberosum subsp. tuberosum ($A_I A_I A_I A_I$)

Dipl. = Dihaploid or diploid species ($A_I A_I$)

Tripl. = Triploid

However, several sterile triploids (second scheme) were also obtained but not doubled.

The produced three-way hybrids OLY x (fen(1) x hje) were all male sterile and not crossable with cv. Olympia. This indicates the disharmony of the parental genomes of these allohexaploids. Hybrids were only obtained after crosses with BGRC 28987 from Denmark. This product was crossable with cv. Olympia, suggesting the elimination of chromosomes, thus returning to the tetraploid level. The late blight resistance of the OLY x ((fen(1) x hje)) x BGRC 28987 hybrid was low in comparison to the nine OLY x (fen(1) x hje) hybrids. Two hypothesis may explain this.
- the high susceptibility of BGRC 28987 (scale 9)
- elimination of the B genome where most likely genes for resistance are located.

The utilization of the LON species present many breeding problems. Considering the high potential value of these species for potato breeding it would be desirable to investigate the cytogenetical background of these breeding problems.

References

Firbas, H., 1961. Beitrag zur Selektion von im Rohzustand nichtdunkelnden Kartoffeln. Z.Pfl.zucht. 46:246-253.

Hawkes, J.G., 1966. Modern taxonomic work on the Solanum species of Mexico and adjacent countries. Am. Potato J. 43:81-103.

Marks, G.E., 1958. Cytogenetic studies in tuberous Solanum species II. A synthesis of Solanum x vallis-mexici Juz. New Phytol. 57:300-310.

Niederhauser, J.S. and W.R. Mills, 1953. Resistance of Solanum species to Phytophthora infestans in Mexico. Phytopathology 43:456-457.

Radcliffe, E.B., F.I. Lauer, M.H. Lee and D.P. Robinson, 1981. Evaluation of the United States Potato Collection for resistance to green peach aphid and potato aphid. Minn. Agr. Exp. Sta. Tech. Bull. 331, 41 pp.

Soest van, L.J.M., 1983. Interspecific hybridization with the allotetraploid tuber-bearing Solanum species S.hjertingii and S.fendleri. Genetika 15:257-268.

Soest van, L.J.M., Barbel Schober and M.F. Tazelaar, 1984. Race non-specific resistance to late blight in tuber-bearing Solanum species and its geographical distribution. Potato Research 27:393-411.

Stelzner, G., 1943. Uber die Fertilitatsverhaltnisse bei Bastardierungen zwischen der frostfesten Wildkartoffel Solanum acaule Bitt. und der Kulturkartoffel Solanum tuberosum L. Der Zuchter 15:143-144.

Toxopeus, H.J., 1964. Treasure-digging for blight resistance in potatoes. Euphytica 13:206-222.

Umaerus, V. and M. Stalhammer, 1969. Studies on field resistance to Phytophthora infestans - 3. Screening of Solanum species for field resistance to P.infestans. Z.Pfl.zucht. 62:6-15.

Woodwards, L. and M.T. Jackson, 1984. The lack of enzymic browning in wild potato species, Series Longipedicellata, and their Crossability with Solanum tuberosum. Z.Pfl.zucht. 94:278-287.

STEROIDAL ALKALOID COMPOSITION OF TUBERS OF EXOTIC SOLANUM SPECIES OF VALUE IN POTATO BREEDING DETERMINED BY HIGH-RESOLUTION GASCHROMATOGRAPHY

W.M.J.van Gelder and H.H.Jonker

Summary

The total steroidal alkaloid composition of tubers from wild and primitive Solanum species was determined with a new method using two-phase hydrolysis and high-resolution gas chromatography. High levels of different types of glycoalkaloids were found in several exotic species. Novel information on the glycoalkaloid composition of Solanum species is presented.

Introduction

In potato breeding, wild and primitive Solanum species are being used to introduce resistance to disease, pests or cold. Together with these traits hazardous levels of solanidine glycosides or other more toxic Solanum glycoalkaloids may be transmitted from exotic species to hybrid progeny. Consequently, when alien germplasm is used in a potato breeding programme, determination of the total steroidal alkaloid composition of parents is of extreme importance. When hazardous levels of glycoalkaloids or teratogenic compounds such as solasodine are present, monitoring of hybrid progeny is necessary too.

Aim of this study

The development of a cromprehensive and efficient method for determining the total steroidal alkaloid composition of Solanum species. Results of a preliminary survey of the alkaloid composition of tubers from some Solanum species which are being used in modern potato breeding are also presented.

Results and discussion

Figure 1 shows the separation of solanthrene, 5-α-cholestane (internal standard), solanidine, demissidine, solasodiene, solasodine and tomatidine by high-resolution gas chromatography of a tuber extract from S.tuberosum cv. Elkana, spiked with 1 mg of solanidine, demissidine, solasodiene, solasodine and tomatidine per ml of extract. Chromatographic conditions were as folows. Column: fused-silica WCOT 50 m x 0.22 mm I.D.; CP-Sil 5 film thickness 0.12 μm. Carrier gas: helium, linear velocity u=24.3 cm/sec. Column temperature: 280°C for 28 min, then increased at 8°C/min to a final temperature of 320°C for 5 min. Injection volume: 1μ 1; split ratio 1:100. Attenuation: 2^{6}. Injector temperature: 325°C. Detector temperature: 350°C.
All compounds were well separated, the resolution of two consecutive peaks was always better than R=1.5 (baseline separation). No compounds interfering with the resolution of the alkaloids were present in the potato tuber extract (for further details see van Gelder, 1984 and 1985).

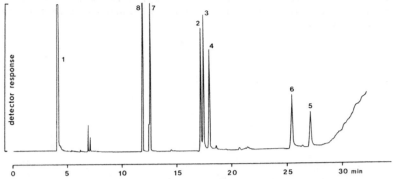

Fig.1. Separation of solanidine (2), demissidine (3), solasodiene (4),
tomatidine (5), solasodine (6), solanthrene (7) and 5- - cholestane
(internal standard; 8) Tuber extract from S.tuberosum cv. Elkana spiked
with components 2,3,4,5,6 and 8.

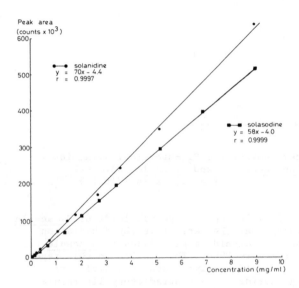

Fig.2. Calibration curves for FID-response of solanidine and solasodine

Figure 2 shows calibration curves for FID-response constructed for
solanidine and solasodine. Calibrations showed linearity over a wide
concentration range, respectively 0.04 - 9.0 mg/ml for solanidine,
corresponding to 5.2 - 1180 mg solanidine glycosides per kg fresh potato
and 0.06 -9.0 mg/ml for solasodine which corresponds to 7.7 - 1154 mg
solasodine glycosides per kg fresh potato.

Chromatographic traces of the steriodal alkaloid composition of tubers
from S.tuberosum ssp. andigena and from S.acaule are shown in figure 3.
Contents of solanidine glycosides calculated from the contents of the

aglycones, varied for different accessions of ssp.andigena (Table 1).
This suggests that intra-specific variation for glycoalkaloid content in
ssp. andigena may exist.

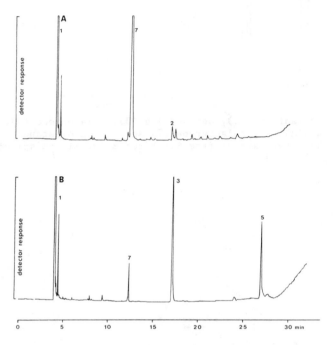

Fig.3. Chromatographic traces showing the C_{27}-steroidal alkaloids from
tubers of S.tuberosum ssp. andigena (A) and S.acaule (B). Peak
indentification: 1, solvent; 2, solanidine; 3, demissidine; 5,
tomatidine; 7, solanthrene.

The chromatographic trace in fig. 3B showing the alkaloids of S. acaule
reveals the presence of solanthrene, the derivative (by dehydration) of
solanidine. On the occurrence of solanidine glycosides in S.acaule the
literature is contradictory (Gregory 1984). The results in this report
prove that S.acaule contains solanidine glycosides in addition to
demissidine and tomatidine glycosides. The contradictory literature
reports can be ascribed to inadequacy of chemical techniques used in
those studies or to intra-specific variation for the presence of
solanidine glycosides in S. acaule.

Table 1. Glycoalkaloid composition of tubers of <u>Solanum</u> species determined by HRGC.

Species	Glycoalkaloids (mg/kg fresh tuber)			
	solanidine glycosides	demissidine glycosides	tomatidine glycosides	unidentif. peaks
S.commersonii 1	990	590	-	several
S.commersonii 2	1980	210	-	several
S.acaule	120	340	630	minor
S.andigena 1	50	-	-	-
S.andigena 2	1130	-	-	minor
S.phureja	24	-	-	-
S.demissum	-	630	780	-

Table 1 shows the tuber glycoalkaloid composition of <u>Solanum</u> species of potential value for plant breeding. Very high levels of several glycoalkaloids were also found in <u>S.commersonii</u> and in <u>S.demissum</u>. In <u>S.commersonii</u> unidentified peaks were found. Work is in progress to investigate wether these peaks represent <u>Solanum</u> alkaloids.

References

Gelder, W.M.J.van, 1984. A new hydrolysis technique for steroid glycoalkaloids with unstable aglycones from <u>Solanum</u> spp. J.Sci. Food Agric.35: 487-494.
Gelder, W.M.J.van, 1985. Determination of the total C27-steroidal alkaloid composition of <u>Solanum</u> species by high-resolution gas chromatography. J.Chromatogr. 331: 285-293.
Greorgy, P., 1984. Glycoalkaloid composition of potatoes: Diversity and biological implications. Am. Potato J. 61: 115-122.

DEVELOPMENT OF PARENTAL LINES FOR POTATO BREEDING AND ITS UTILIZATION IN POLAND

K.M. Swiezyński, B. Czech, M. Dziewońska, J. Olejniczak, E. Ratuszniak, M. Sieczka

Institute for Potato Research, Młochów, 05-832 Rozalin, Poland

Parental lines are being developed in the Institute for Potato Research. They are available to all potato breeders in Poland. In the period 1967-1985 following types of parental lines have been supplied to the breeders:

Characters in which parental lines were outstanding	Number of samples supplied:		
	1967-79	1980-85	total
High table quality	117	6	123
" + XY	4	–	4
" + N	11	–	11
First early tuber formation	48	–	48
" + XY	46	3	49
" + XYS	–	1	1
High starch content in the tubers	78	–	78
" + XY	93	17	110
" + XYSPh	–	18	18
" + YPhN	–	3	3
Drought resistance	5	8	13
" + XY	70	16	86
" + XYS	–	10	10
" + XYSPh	–	24	24
Early with satsifact. starch content			
" + res. to 1-2 vir. or Ph	50	5	55
" + res. to 1-2 vir. and N	–	28	28
" + XYSPh	50	77	127
" + XYSPhN	–	6	6
High starch content with satisfact. protein content	86	–	86
" + storability or N	4	1	5
Multiple resistance to 4 viruses: XYSL or XYML	–	14	14
Others	237	206	443
Total	899	443	1342

Explanations: X, Y, S, M, L - resistance to respective potato viruses (L for leafroll); Ph - resistance to late blight; N - resistance to nematodes; storability - some resistance to storage diseases.

At present about 30% potato breeding materials in Poland and 3 cultivars: Brda, Bzura and Poprad originate from parental lines supplied by the Institute for Potato Research.